A Longstanding Attraction

Alberto P. Guimarães

A Longstanding Attraction

The History and Modern Uses of Magnetism

Second Edition 2025

Alberto P. Guimarães
CBPF
Rio de Janeiro, Brazil

ISBN 978-3-032-02005-5 ISBN 978-3-032-02006-2 (eBook)
https://doi.org/10.1007/978-3-032-02006-2

General Interest
1st edition: © Author 2005

This Springer imprint is published by the registered company Springer Nature Switzerland AG
The registered company address is: Gewerbestrasse 11, 6330 Cham, Switzerland

Dedicated to Silvia Bregman

Foreword

Magnetism is one of the oldest and most studied areas of physics, the field of natural science that investigates the constituents of matter, their interactions, and behavior through space and time. A scientist who specializes in the field of physics is called a physicist. As Alberto P. Guimarães describes in detail in this book, magnetism was first observed and reported in the ancient world when it was noticed that lodestones, naturally magnetized pieces of the mineral magnetite, could attract pieces of iron at a distance. This intriguing property of matter was used in ancient China to make a compass, a device that uses an elongated piece of lodestone that points along the South–North direction of the Earth to help travelers in long-distance navigation.

Studies of magnetism were conducted in other Eastern countries and Greece for several centuries BC. They gained new impetus in the Middle Ages with the development of science in Western Europe. A few books on the subject were written by French and British scientists. However, the first scientific book on magnetism was *De Magnete*, written in Latin by the British physician William Gilbert, published in 1600. In *De Magnete*, Gilbert describes many experiments with his model Earth, called the terrella. From his experiments, he concluded that the Earth was itself magnetic and that this was the reason compasses pointed North, whereas, previously, some believed that the pole star Polaris or a sizeable magnetic island on the North pole attracted the compass. *De Magnete* is considered one of the most significant books in the history of sciences that influenced many scientists on the European continent, notably two important almost contemporaries of Gilbert, Johannes Kepler, and Galileo Galilei.

Research in magnetism and electricity in Europe in the nineteenth century revealed that phenomena in the two areas are related, namely an electric current can create a magnetic field and, conversely, a magnetic flux can create an electric current. These discoveries led to the invention of electric motors and generators, which became essential devices for humanity. Also, they gave birth to the field of electromagnetism, governed by the beautiful equations formulated by the Scottish physicist James Clerk Maxwell at the end of the nineteenth century. However, as Alberto very well describes, only in the twentieth century, with the development of quantum mechanics, the microscopic origins of the magnetic properties of matter became understood.

Designed primarily for the general public, *The Magic of the Magnet: A Short History of Magnetism* will also interest students and scientists in other areas of science. It provides an excellent presentation of the history of magnetism, the evolution of its understanding, and its applications over the centuries up to the present day.

Physics
Universidade Federal de Pernambuco
Recife, Brazil

Sergio M. Rezende

Preface

Many readers recall the amazement they felt when they first saw a magnet. Why does the magnet attract iron objects? This question has been asked many times for thousands of years.

The relationship between magnetic phenomena and electrical effects was another challenge that sparked many investigations, leading to fundamental discoveries and a wide range of applications.

The explanation of the magnetic effects was only discovered in the modern era. Finally, the understanding of magnetic phenomena and the development of magnetic materials have led to large-scale applications of magnetism in several areas, particularly as the technology that guarantees the long-term preservation of data relevant to all areas of human activity.

This text is an expanded and thoroughly updated version of an early publication by the author, *From Lodestone to Supermagnets: Understanding Magnetic Phenomena*, Wiley-VCH, Weinheim (2005).

The author is indebted to Wiley-VCH GmbH for reversing the rights of the work mentioned above. The author is also indebted to the Bakken Museum of Minneapolis, US, for a Visiting Research Fellowship in 2001 to consult their rich collection and search for iconography on magnetism.

The work of the reviewer, Max H. Rumjanek, is acknowledged. I would also like to thank the comments of those who read parts of the manuscript: N. Alves, G. O. Corrêa, N. Countryman, D. A. S. E. Cruz, D. Franceschini, F. L. Freire, F. Garcia, M. Z. P. Guimarães, R. F. N. Guimarães, R. Z. P. Guimarães, R. Lent, A. Mello, I. S. Oliveira, E. Passamani, S. M. Rezende, A. M. P. Silva, J. P. Sinnecker, and C. L. Vieira.

Rio de Janeiro, Brazil — Alberto P. Guimarães

Contents

About the Author

Alberto P. Guimarães is an Emeritus Researcher at the Centro Brasileiro de Pesquisas Físicas (CBPF) in Rio de Janeiro and is a specialist in magnetism. He obtained his Ph.D. at the University of Manchester, England. He is the author of the books, including *Principles of Nanomagnetism*, Springer, 2nd edition (2017), *From Lodestone to Supermagnets: Understanding Magnetic Phenomena*, Wiley-VCH (2005), and *Magnetism and Magnetic Resonance in Solids*, John Wiley (1998).

1

Introduction: The Magic of the Magnet

A wonder of such nature I experienced as a child of 4 or 5 years, when my father showed me a compass. That this needle behaved in such a determined way did not at all fit into the nature of events, which could find a place in the unconscious world of concepts (effect connected with direct "touch"). I can still remember—or at least believe I can remember—that this experience made a deep and lasting impression upon me. Something deeply hidden had to be behind things.
Albert Einstein, Autobiographical Notes [1].

Summary The phenomenon of magnetic attraction by the lodestone has been known for thousands of years. The compass was the first practical application of magnetism, used in navigation in China in the tenth century AD. Another effect known for centuries was the attraction and repulsion between charged objects. In the nineteenth century, researchers started investigating the connection between these two phenomena. The decisive experiment was performed by Hans Oersted (1777–1851), who found that the needle of a compass moved when an electric current flowed in a nearby wire. Michael Faraday (1791–1867), in the same century, revealed the symmetric effect, i.e., magnets inducing an electric current. Many applications resulted from these discoveries, with the creation of new techniques and equipment. Medicine benefited, e.g., with the use of nuclear magnetic resonance or the magnetic properties of hemoglobin for images and the therapeutic and diagnostic use of magnetic nanoparticles. Only in the twentieth century, with the birth of quantum mechanics, was the physics of magnetism understood. How could one predict that this strange phenomenon would be used to preserve the memories of humankind? Today, information created at an accelerating rate is stored in magnetic tapes and magnetic hard disks.

A. P. Guimarães, *A Longstanding Attraction*, https://doi.org/10.1007/978-3-032-02006-2_1

The intriguing phenomenon of magnetic attraction has been known for thousands of years, as demonstrated by references in texts from Mesopotamia. The first practical application found for magnetism was the magnetic compass, used in navigation in China in the tenth century AD. Another effect known for centuries involved charged objects; they also show attraction and repulsion. Only in the nineteenth century did researchers investigate the possible connection between these two phenomena. The Danish scientist Hans Oersted (1777–1851) performed the most decisive experiment, and he found that the compass needle moved when an electric current flowed through a nearby wire. Investigations conducted by the Englishman Michael Faraday (1791–1867) in the same century revealed the symmetric effect, i.e., magnets may induce the circulation of an electric charge. Many essential applications resulted from these experiments, with the creation of new machines and equipment. The origin of magnetic phenomena was explained in the twentieth century with the birth of quantum mechanics. How would one have predicted that knowledge of this weird phenomenon would lead to a new way of preserving the memories of humankind? This point is where we stand today, with magnetic hard disks and magnetic tapes storing information produced at an explosive rate in our era.

Who has not felt intrigued when faced with a magnet attracting a piece of iron? For how long has the phenomenon of magnetic attraction been known?

The invention of writing, which led to the creation of the cuneiform language around 3000 BC, allowed the preservation of early references to magnetism. Texts from Mesopotamia of the second millennium BC mention materials with this property [2]. These materials are minerals like magnetite, a naturally occurring iron oxide of formula Fe_3O_4. The name magnetite comes from the region of Magnesia, near Lydia, in Asia Minor, in present-day Turkey.

The beginning of science is generally associated with the first Greek philosophers, or "lovers of wisdom," who were active from the sixth century BC [3]. The first name recorded is Thales of Miletus (c. 640 BC–546 BC), whose works have not been preserved. The great philosopher Aristotle (384 BC–322 BC), who lived some two centuries later, mentioned that the magnet drew the attention of Thales.

Some of the earliest references to the lodestone come from the East, mainly from China, where it was called *the loving stone* [2]. The first written record was published in 240 BC. The first application of magnets—the compass—originated in China, as did its use for navigation, possibly in the tenth century AD. In many technical fields, the Chinese were ahead of Europeans until the sixteenth century. In the first century AD, lodestones mounted on a pin were common.

The description of a compass in China (about AD 1080) antedated by 100 years the first European records. However, a new paradigm in the description of Nature (by the seventeenth century) arose only in Western Europe, the so-called scientific revolution.

The Arab world, at its apogee (mid-eighth century AD), extended from Spain to India. Besides their scientific contributions, the Arabs played an important role in preserving and spreading the works of the Greek thinkers in the West. Their first text on the compass was the *Jami al-Hikayat* (around 1232). Arab mariners may have used the compass as early as the eleventh century (Fig. 1.1).

The compass appeared in Europe about AD 1150. In 1269, in the siege of Lucera, Italy, during the campaign of King Charles of Anjou (Charles I) (1226–1285), the Frenchman Pierre de Marincourt (or Petrus Peregrinus) (born c. 1220) wrote the first European work devoted to the magnetism of the lodestone, the "*Letter on the Magnet*" [4]. He announced: "…this stone bears in itself the likeness of the heavens."

However, the most important treatise on magnetism in this period was the "*De Magnete*" [5] (1600) by William Gilbert (1544–1603), a physician at the court of Queen Elizabeth I (1533–1603). A significant work in the history of the sciences, it is also the first treatise on electricity. Gilbert's use of the experimental method and his critical attitude toward the classics were very innovative.

The most baffling aspect of magnetism is how one object acts on another without a material link between them. In the sixteenth and seventeenth centuries, the idea of action at a distance gained increasing acceptance, culminating with the gravitational theory proposed by Isaac Newton (1642–1727).

Fig. 1.1 Compass needle mounted on a pivot ("dry mount")

Newton considered that the Earth attracted the Moon, its motion deviating from a straight line due to "falling" to the Earth, like any terrestrial object.

Until the eighteenth century, it was not known that electrification by friction and the transport of current through conductors involved the same charge carriers. The hypothesis of one single electrical fluid was made by Benjamin Franklin (1706–1790).

Pliny, the Elder (23–79 AD), acknowledged the similarity between the attraction of amber (a fossilized tree resin) and that of the magnet [6]; this suggested some link between the two phenomena. In 1820, Hans Christian Oersted (1777–1851) proved it, observing that an electric current created a field that moved a compass needle.[1]

André-Marie Ampère (1775–1836) showed that a coil of copper wire, if free to turn, behaved as a compass needle. Michael Faraday (1791–1867) obtained an electric current with a magnet moving in and out of a coil, inventing the first generator. James Clerk Maxwell (1831–1879) assumed that light and electrical disturbances were the same: propagating electromagnetic waves.

The idea that all matter is constituted of small particles (atoms) was first proposed by Democritus (c. 460–c. 370 BC), a Greek thinker. This hypothesis was confirmed much later, as the English researcher John Dalton (1766–1844) studied the proportion of the elements required to form a compound, e.g., the volumes of hydrogen and oxygen required to form water (of formula H_2O). Only in the twentieth century did the New Zealander Ernest Rutherford (1871–1937) determine that atoms have a tiny nucleus, around which the electrons turn. The Danish physicist Niels Bohr (1885–1962) published in 1913 the first model that described the atoms. The German Max Planck (1858–1947) proposed that the atoms absorbed and irradiated energy in packets or "quanta." The German-American Albert Einstein (1879–1955) later advanced the idea that light and radio waves carried discrete packets of energy or quanta.

The ideas of Planck, Einstein, and others defined a new physics, subsequently known as quantum mechanics. In this conception, all matter, including the particles, presents wavelike properties.

One of the triumphs of this new theory is that it explained the existence of magnetism, which rests on a quantum interaction ('exchange') between electrons.

Medicine was organized in Egypt almost 3000 years before our era. Medicine developed in Greece through the influence of the Egyptians. At

[1] Oersted, H. C. (1830). Thermo-electricity, in *Edinburgh Encyclopaedia* (1830), (XVIII, pp. 573-589), quoted by Williams [7].

about the same time, it also developed in China and India. There are very old records of magnets employed by physicians, for example, to extract swallowed pieces of iron or pieces of iron in the eyes.

Medical techniques developed over the years, use the magnetic properties of nuclei of some elements to obtain images of the different organs in magnetic resonance imaging (MRI) and functional MRI (fMRI). The magnetic properties of hemoglobin are also used to obtain images of the organs. Microscopic magnetic particles are used to destroy tumors.

The written word revolutionized the diffusion and accumulation of knowledge, leading to the invention of the book. One estimates that the text of an average book contains about one million bytes of information (1 megabyte, or 1 MB).

Data are recorded in a magnetic recording apparatus as tiny magnetized regions on a recording medium [8].Credit cards and ATM cards also use magnetic memory.

In the first magnetic tape system (1951), the data density was about 100 million times smaller than today's. Magnetic hard disks (HDDs) had in 1956 an areal density of 2 kilobytes per square inch, or 2 kB/in^2, whereas their present density is of the order of 1,000,000,000,000 bytes/in^2 = 1 terabyte/in^2 = 1 TB/in^2.

HDDs provide direct access to data, i.e., a given point on the disk can be reached in a very short time; on tapes, a much longer time is needed to read the information of interest, although tapes are cheaper and last longer.

HDDs have recently incorporated new technology known as energy-assisted magnetic recording (EAMR); also, new HDDs may use superposed recording tracks.

The total amount of information produced worldwide per year is expected to reach more than 2100 zettabytes (2.1×10^{24} bytes) in 2035 [9]. A pile of books containing this information would be millions and millions of kilometers high! This fact illustrates the information explosion, one of the characteristic aspects of our era.

In 2025, one expects that the information in all the data centers where information is stored—mostly in magnetic media—will reach 2×10^{23} bytes. The electricity used by these centers in 2021 was around 0.9–1.3% of global electricity demand.

Magnetic storage is, therefore, the main form of preserving information for humankind.

References

1. Einstein, A. (1969). Autobiographical notes. In P. A. Schilpp (Ed.), *Albert Einstein: Philosopher-scientist* (Vol. I, p. 9). Cambridge University Press.
2. Needham, J. (1972). *Science and civilisation in China* (Vol. 4, part I). Cambridge University Press.
3. Taton, R. (Ed.). (1966). *Histoire Générale des Sciences* (Vol. I, p. 204). Presses Universitaires de France.
4. Grant, E. (1980). Peter Peregrinus. In C. C. Gillispie (Ed.), *Dictionary of scientific biographies*. Charles Scribner's Sons.
5. Gilbert, W. (1978). *De Magnete* (Great books) (Vol. 28). Encyclopaedia Britannica. (Original work published 1600).
6. Pliny the Elder. (1991). *Natural history: A selection*, Book XXVI, no. 127 (p. 370). Penguin Books.
7. Williams, L. P. (1980). Oersted. In C. C. Gillispie (Ed.), *Dictionary of scientific biographies*. Charles Scribner's Sons.
8. Piramanayagam, S. N., & Chong, T. C. (Eds.). (2012). *Developments in data storage: Materials perspective*. John Wiley.
9. Armstrong, M. (2019). *Global data creation is about to explode*. https://www.statista.com/chart/17727/global-data-creation-forecasts/

Further Reading

Blundell, S. (2012). *Magnetism, a very short introduction*. Oxford University Press.

Gillispie, C. C. (Ed.) (1970–1980). *Dictionary of scientific biography*. Charles Scribner's Sons, 16 vols; Holmes, F. L. (Ed.) (1990). *Supplement II*, 2 vols.

Guimarães, A. P. (2005). *Lodestone to supermagnets: Understanding magnetic phenomena*. Wiley-VCH.

Keith, S. T., & Quédec, P. (1992). Magnetism and magnetic materials. In L. Hoddeson, E. Braun, J. Teichmannn, & S. Weart (Eds.), *Out of the crystal maze, chapters from the history of solid-state physics*. Oxford University Press.

Livingston, J. D. (1996). *Driving force, the natural magic of magnets* (Vol. 64). Harvard University Press.

Needham, J. (1972). *Science and civilisation in China* (Vol. 4, part I). Cambridge University Press.

Pumfrey, S. (2002). *Latitude & the magnetic earth*. Icon Books.

2

The Stone with a Soul: Stones That Attract

The secret of magnets, now explain that to me!
There is no greater secret, except love and hate.
*Johann Wolfgang von Goethe (*Gott, Gemüt und Welt*) ("Magnetes Geheimnis, erkläre mir das! Kein grösser Geheimnis als Lieb und Hass." Johann Wolfgang von Goethe (1749–1832), "Gott, Gemüt und Welt" (1992).* A Dictionary of Scientific Quotations, *Alan L. Mackay (Compiler), (2nd ed., p. 105). Institute of Physics.)*

Summary Who has not felt intrigued when faced with a magnet attracting a piece of iron? For how long has the phenomenon of magnetic attraction been known? The invention of writing, which led to the creation of the cuneiform language around 3000 BC, allowed the preservation of early references to magnets. Texts from Mesopotamia of the second millennium BC mention a "grasping hematite" or "hematite that seizes." The first materials known to have this property are minerals like magnetite, a naturally occurring iron oxide of formula Fe_3O_4. The name magnetite comes from the region of Magnesia in Asia Minor, present-day Turkey. The beginning of science is generally associated with the first Greek philosophers, or "lovers of wisdom," active in the sixth century BC. The first name recorded is Thales of Miletus (c. 640–546 BC), whose works have not been preserved. The great philosopher Aristotle (384–322 BC), who lived some two centuries later, attributed to Thales the remark that the magnet had a soul. The first application of magnetism was the creation of the compass in China. It was used for navigation, possibly in the tenth century AD.

A. P. Guimarães, *A Longstanding Attraction*, https://doi.org/10.1007/978-3-032-02006-2_2

It is not easy to pinpoint when magnets' property of attracting some materials, such as iron, was discovered. The observation of magnets has accompanied humankind for more than 3000 years, for these wondrous objects were known before the first millennium BC. In ancient Mesopotamia, iron oxides were used in weights from the late third millennium BC [1]. Different iron ores, including the mineral magnetite, were used in the same region to make seals ('cylinder seals') since 2000 BC.

The earliest indication that magnetic phenomena were known in the ancient world is the fact that the magnetic mineral magnetite, the naturally occurring magnetic stone, was referred to in Mesopotamia as "grasping hematite" or "hematite that seizes" (*shadânu sabitu*) [2–4]. This term is used in a tablet with a list of commodities[1] from the first half of the second millennium BC, indicating that this remarkable property of some iron ores had been observed very early. The expression "grasping hematite" is also found in the 16th tablet that is part of a series called *Har-ra hubullu*, containing Sumerian words and the Akkadian equivalent from the first millennium BC.[2] The phrase "living hematite" is also recorded.

The first known records of the properties of the magnet were made in Greece. The Greek philosopher Thales of Miletus, who lived in the sixth century BC, was said to have considered that the magnet had a soul.

The origin of the name magnet, according to the Roman poet-philosopher Lucretius Lucretius (Titus Lucretius Carus) (c. 99–c. 55 BC), writing in the first century BC in the didactic poem *De Rerum Natura* ("On the Nature of Things"), was in the province of Magnesia, in Thessaly, northern Greece. Lucretius wrote [5]: "Next in order I will proceed to discuss by what law of nature it comes to pass that iron can be attracted by that stone which the Greeks call the Magnet from the name of its native place, because it has its origin within the bounds of the country of the Magnesians." Inhabitants of Thessaly colonized Asia Minor, where there are two provinces named Magnesia. Since the lodestone is referred to as "Lydian stone" in some accounts, including one[3] of the Greek poet Sophocles (496–406 BC), one may relate the discovery to the Magnesia in the region of Lydia [7].

A different account is given by Pliny the Elder (Gaius Plinius Secundus) (AD 23–79), a Roman author who wrote the encyclopedic *Natural History* (*Historia Naturalis*), a 37-volume treatise whose influence lasted for some 15 centuries. Pliny compiled thousands of observations on natural phenomena,

[1] J. Huehnergard, private communication (2000).

[2] J. Huehnergard, private communication (2004).

[3] A. C. Pearson, quoted by J. B. Kramer [6].

plants, animals, and places from his own experience and from more than 2000 earlier texts. He was untiring in this task, studying and writing during his whole life. His curiosity finally led to his death as he approached the volcano Vesuvius to observe more closely the eruption of the year AD 79 that destroyed Pompeii.

He mentions the property of magnets of attracting iron. He informs us that according to Nicander (a Greek poet), "it was known as *magnes* after its discoverer, Magnes, who found the mineral on Mount Ida.[4] (…) The discovery is said to have been made when the nails of Magnes' sandals and the tip of his staff stuck to the stone as he was grazing his herds" [8].

It is difficult to establish which of the traditional versions of the origin of the word "magnet" presented above is more reliable. In any case, we know that the first objects to show this amazing property were rocks containing iron oxides, a mineral known as magnetite, basically an iron oxide of formula Fe_3O_4, brown or black, with a metallic luster. The magnetite that behaves as a magnet is known as lodestone, from the word *lode*, an archaic English meaning course.

The knowledge of magnetism and magnetic materials was acquired very slowly since the beginning of science. Knowledge about minerals has been accumulated since prehistoric times, as humankind observed and classified natural materials according to their physical properties, used them, and later tried to tailor them to their practical needs. The accumulation of knowledge accelerated as the first human groups started to settle, abandoning nomadic life, and began to raise the first crops in Mesopotamia, the region between the rivers Tigris and Euphrates in present-day Iraq, and also along the river Nile, in Egypt.

These civilizations learned how to grow wheat and barley, raise cattle and work metals like copper and bronze (an alloy of copper and tin). Iron was initially extracted from metallic meteorites; iron extracted from ore appears in the first half of the third millennium BC [9]. Their way of life required a certain amount of botanical knowledge, knowledge of the cycle of seasons, and the knowledge about the melting points of the metals, their degree of hardness, malleability, and so on.

The development of the reckoning of time was related to the growth of astronomical knowledge. Time was naturally measured in days, the period from sunrise to sunrise (corresponding to the rotation of the Earth around its axis); a longer period was recognized as the year, the interval from one cycle of

---

[4] There is a Mount Ida in Asia Minor (Turkey); in the Illiad Zeus watched the Trojan War from Mount Ida, also known locally as Kaz Dagi.

seasons to the next (corresponding to the time of revolution of the Earth around the Sun). The determination of how many days were contained in a year led in Mesopotamia and Egypt to the introduction of a calendar that incorporated the yearly periodicity of the seasons, important for the agricultural society; it was convenient to start the seasons on the same day every year. This demanded the observation of stars and, in general, the acquisition of astronomical data. The division of the day into hours was not so relevant for life in a primitive society, where the activities were limited from dawn to dusk. Also, measurement of time on this shorter scale involves creating new instruments and solving many technical problems.

The construction of housing and temples and the need to measure land led to the development of units of measurement; a system of weights and measures was created in Mesopotamia as early as 2500 BC, and later also in Egypt. The usual units of length were related to parts of the human body: the cubit, the distance from a man's elbow to the fingertip; the span, the length of a fully stretched hand, from the tip of the thumb to the tip of the little finger; and so on. One of the earliest units of weight used in Mesopotamia was the *mina*; another unit was the *shekel*, equivalent to 1/60 of the *mina*. The *shekel* was equivalent to the weight of 120 or 200 grains of wheat [10].

A giant step in the development of early human society was the invention of writing by the Sumerians, one of the peoples of Mesopotamia. It started as a pictographic representation, where, e.g., a picture of the head of the animal represented an ox and later took a more abstract form in the so-called cuneiform language around 3000 BC. This name arises from the shape of the short strokes (from the Latin cuneus, meaning "wedge"). The next stage in the evolution of writing was the shift from representing things to representing sounds.

Thales and the Beginnings of Greek Science

The beginnings of science are usually placed in Greece, in the sixth century BC. There is a certain degree of arbitrariness in this choice since scientific and technical developments were, of course, also made in Egypt, Babylonia, and China. However, it is generally accepted that the kind of knowledge, or rather the approach towards knowledge itself, of the first investigators[5] in Greece had the seeds of the scientific endeavor.[6]

Why Greece? Why did it not occur in Egypt? It is known that in Egypt, despite the wealth of practical knowledge accumulated, there is no evidence of systematic theorizing on natural phenomena. In Egypt, the identification of the gods to the kings, with the priests being both religious and secular authorities, did not contribute to creating a climate that stimulated free speculation about the universe [13]. In China, on the other hand, law did not guarantee individual rights; customs and a "natural law" prevailed; without the existence of prescribed sanctions for different crimes [14]. It has been speculated that since there was no immediate connection in China between human acts and their legal sanctions, the relationship between the natural phenomena and general laws that governed them was not easily perceived; the concept of laws of nature, therefore, was not valued.[7]

In the Greek world, one of the elements that contributed to the emergence of science was the fact that the gods were not associated with secular power, and as a consequence, they could be more easily removed from the privileged role of major actors that determined the course of natural events. In the early mythical accounts, nature tended to be endowed with the power to evolve from within, a useful characteristic to open the way for the causal explanation of natural phenomena. The creation of the universe, as described in Hesiod's Theogony, is not the act of gods but rather resulted from the action of "relatively abstract entities," such as Chaos, Earth, and Eros [16] (Theogony is an account of the origin and descent of the gods).

Political factors have also been pointed out as relevant for this development: the Greek city-states enjoyed a variety of constitutional forms of organization, among these the novel democratic regime, which allowed the

---

[5] Important ideas for the emergence of scientific thinking: "The four ideas are (1) the Ionian natural philosopher's way of explaining nature, (2) the emphasis on rational argumentation, (3) Aristotle's introduction of the concept of logical validity, and (4) Euclid's axiomatic mathematics" [11].

[6] "To the extent that it is explicative and ontological [i.e., related to the being], science is a creation of the Greek genius—and if one considers these two aspects as essential, one can assert that science was born in Greece" [12].

[7] J. Needham, quoted by P. Acot [15].

development of an atmosphere of free speculation. Indeed, in the case of Athens, the constitution permitted a high level of participation in the political life of the city [17]. The amplitude of the maritime trade, with the exchanges between different cultures, particularly with the Egyptians and the peoples of Mesopotamia, may have worked in favor of expanding the horizons of the first thinkers. A higher level of literacy in Greece, as compared to Egypt and Babylonia, was again a favorable factor [18].

The Greek world in antiquity was constituted of cities scattered on the edges of the Mediterranean Sea (Fig. 2.1); Egyptian civilization was concentrated on the banks of the Nile, relatively isolated from other societies by the desert, east and west of the river. Both in Egypt and in Greece, however, intellectual life could develop only when society spared some of its members from productive activities, allowing them to enjoy free time for leisure, as first pointed out by the philosopher Aristotle: "Hence, when all such inventions were already established, the sciences which do not aim at giving pleasure or at the necessities of life were discovered, and first in places where men first began to have leisure. This is why the mathematical arts were founded in Egypt; for there the priestly caste was allowed to be at leisure" [19]. And furthermore, "(.) for it was when almost all the necessities of life and the things that make for comfort and recreation had been secured, that such knowledge began to be sought."

Around 1400 BC, the first writing system based on the sounds of the words was invented in Greece. This evolved some centuries later to become an alphabetic system, where syllables were formed with consonants and vowels. Many hundreds of clay tablets with inscriptions were found, mostly in Crete, written in the period 1450–1200 BCE in a language known as Linear B, identified as early Greek [20].

About 750 BC [21], there appeared the Iliad and the Odyssey, epic poems attributed to Homer (eighth century BC), a poet-singer who would have lived

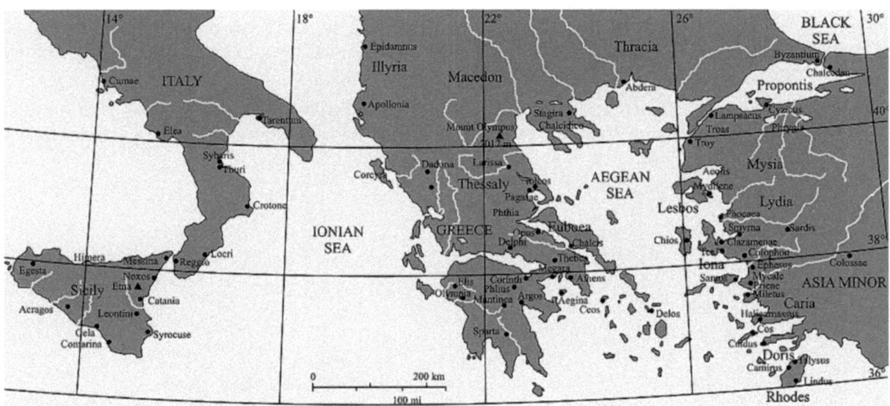

Fig. 2.1 The Greek world

possibly on Chios, an island on the coast of Asia Minor or Anatolia, the Asian portion of Turkey. Initially transmitted through oral tradition, these masterpieces of Greek literature were centered on the adventures of the legendary characters Achilles and Ulysses (Odysseus); in the Iliad, the history of the Trojan War is told (Ilion is one of the names of Troy). Besides their great literary value, these works are important as they reveal a certain knowledge of natural phenomena and of medical and agricultural practices prevalent in Greece at that time [22]. The Iliad and the Odyssey helped to consolidate the Greek language and became part of the education of every literate Greek for many centuries. Two other poems, "Theogony" and "Works and Days," are attributed to Hesiod, who may have lived in Ascra, in Boeotia, mainland Greece, around 700 BC. Herodotus, the Greek historian and geographer who lived in the fifth century BC, expressed the importance of Hesiod and Homer in shaping Greek culture with the following words: "It is they who created a theogony for the Greeks, gave the gods their names, distributed their privileges and skills, and described their appearance."[8] Or, in the words of Xenophanes, a philosopher and religious thinker from Colophon, in Ionia (the central part of the coast of Asiatic Turkey), who lived in the sixth to fifth century BC: "What all men learn is shaped by Homer from the beginning" [23]. The poems of Hesiod and Homer, that mingled myth with practical knowledge, mark the transition of Greece into a new intellectual era.

This era opened in the sixth century BC, as some inquisitive men start to ask new questions about the world that surrounded them and, furthermore, try to search for answers that did not contain the usual elements of lore and mythology; this meant trying to find natural causes for the natural phenomena. In these first bold endeavors lie the origins of scientific thought. These men were the philosophers, or lovers of wisdom, in Greek; Aristotle would later on refer to them as *physikoi*, from *physis*, nature. In their standpoint, they avoided the attitude common in pre-scientific societies of mistaking "association of ideas in their minds" for "causal relations between things," in the words of the nineteenth century British anthropologist James Frazer.[9] First of all, they began to view the world not as some disordered, or arbitrary, aggregate of things or sequence of events, but as an object of knowledge and explanation. The word *kosmos*, from a verb meaning "to order", was used to describe the universe; the very fact that a word was required to designate the totality of things represents, in itself, an important move in this new direction. Their attitude is summed up in the words of W. K. C. Guthrie [25], author of "*A History of Greek Philosophy*": "Philosophy started in the faith that beneath this

[8] Herodotus, Hdt. II, 53, quoted by W. K. C. Guthrie [23].

[9] J. Frazer, *The Golden Bough*, London, 1890, quoted by R. S. Brumbaugh [24].

apparent chaos there exists a hidden permanence and unity, discernible, if not by sense, then by the mind."

It is very difficult for us nowadays to apprehend the magnitude of the change in worldview that the attitude of the early philosophers entails since so much of their contribution has become a tacit part of our intellectual inheritance.[10] We are used to searching for natural causes of phenomena without having to make the conscious choice between a mythical and a scientific standpoint; under most circumstances, the choice between *mythos* and *logos* is immediate.

The first philosophical ideas appeared in the eastern and western frontiers of the Greek world (Fig. 2.1): in the East, along the cities of the coast and in the islands of Asia Minor, and in the West, in Sicily and mainland Italy. Very little original written work or even fragments of works of the early philosophers have survived to our days. Most of their ideas reached us through the writings of commentators or historians who lived centuries after them and who did not have access to the original manuscripts; to make things worse, even these later accounts were preserved in the form of copies of copies of texts [27].

The first of these thinkers is Thales of Miletus (c. 640–546 BC), who is regarded as the founder of Greek science and philosophy. He made the daring move of proposing that there was something common to the entire universe. In this first unified view of the universe, he posited, according to Aristotle,[11] that "all things are water." The choice of water as the element that formed everything in the universe may have arisen from the fact that water can present itself as a solid, as a liquid, and as a gas. This choice for the prime element of the universe may also have resulted naturally from the importance of water in the life of animals and plants; the relevance of the sea for the economic activity of the peoples in the region may have also played some role. This idea had antecedents, since many early myths involved sea and water deities, and Babylonian cosmology around perhaps 2000 BC already assumed a primacy of water [29]. It was also part of these myths that the Earth itself was regarded as floating on an infinite pool of water.

[10] The philosopher Karl Popper places the originality of the first philosophers not so much in the fact that they abandon the myths, but in their critical attitude: they substitute the "traditional preservation of the dogma", for a "tradition of criticizing theories" [26].

[11] "Most of the first philosophers thought that principles in the form of matter were the only principles of all things. (…) Thales, the founder of this kind of philosophy, says that it is water (that is why he declares that the earth rests on water)" [28].

"It was Thales who first conceived the principle of explaining the multiplicity of phenomena by a small number of hypotheses".[12] He proposed the first vision of the universe that lacked mythological elements, gods or demons, as essential constituents. He had a reputation as a mathematician, and he was the first man to be considered responsible for specific mathematical discoveries [30]. However, some of his alleged scientific exploits are not generally accepted today, such as the famous prediction of a solar eclipse [31–33] in 585 BC; it is argued that there simply was not sufficient astronomical knowledge at the time to allow such a prediction.

Two other thinkers constitute, with Thales, the School of Miletus: they are Anaximander (610–550 BC) and Anaximenes (550–486 BC). Anaximander elaborates on the worldview of Thales, assuming a single element as the principle of everything, but in his view, this was an abstract element, the *apeiron*, or the infinite. This choice represents a more sophisticated attempt at the comprehension of the principle, or *physis*, of the universe, since it shifts from known material substances to a pure abstraction.

Simplicius, a commentator and philosopher of the Neoplatonist school, who was born in the second half of the fifth century AD, wrote[13] of Anaximander: "Of those who hold that the first principle is one, moving and infinite, Anaximander, son of Praxiades, a Milesian, who was a successor and pupil of Thales, said that the infinite is principle and element of the things that exist. He was the first to introduce the word "principle." He says that it is neither water nor any other of the so-called elements, but some different infinite nature, from which all the heavens and the worlds in them come into being. And the things from which existing things come into being are also the things into which they are destroyed, in accordance with what must be."

The philosopher Anaximenes, in his turn, chooses air (or breath) as the prime element. He proposes mechanisms of condensation and rarefaction as the processes of formation of the variety of qualities and material aspects of the universe. Condensation of air produces cold; further condensation creates wind, clouds, rain, earth, and rock. In an ancient text falsely attributed to Plutarch,[14] one reads: "Anaximenes, son of Eurystratus, a Milesian, asserted that air is the first principle of the things that exist; for everything comes into being from air and is resolved again into it. For example, *our souls*, he says, *being air, hold us together, and breath and air contain the whole world* ('air' and 'breath' are used synonymously)."

[12] S. Sambursky, *The Physical Word of the Greeks* (Trans. M. Dagu, London 1956, p. 7), quoted by Robert S. Brumbaugh [24].

[13] Simplicius, "Commentary on the Physics" 24.13-25, in Barnes [34].

[14] [Plutarch], "On the Scientific Beliefs of the Philosophers," 876 AB, quoted by J. Barnes [35].

Another great philosopher of importance in the history of science was Pythagoras; he was born on Samos, in Ionia, on the island closest to the continent, probably about the year 570 BC. To flee from the tyranny of Polycrates, he migrated to Croton, in Southern Italy, where he established his school. He was then 40. His presence in Croton coincided with a period of increase in the influence of this city-state. At the end of the sixth century, after a rebellion in Croton, Pythagoras migrated again, ending his days in Metapontum, also in Italy, in the Gulf of Tarentum.

Besides being a philosopher and a political leader, Pythagoras created a religious sect; the facts of his life are entangled with many myths concerning his person. There is no certainty that Pythagoras left written works; a veil of secrecy covered his religious society, where members were obliged to take a 5-year vow of silence [36]. His philosophical work was apparently accessory to his religious interests [37]: "What we may safely say is that for Pythagoras religious and moral motives were dominant, so that his philosophical inquiries were destined from the start to support a particular conception of the best life and fulfill certain spiritual aspirations."

For the Pythagoreans, numbers had an independent existence and were more important than the natural phenomena; these were important insofar as they reflected the preeminence of numbers [38]. This supreme position of the numbers was described by the philosopher Aristoxenus (fl. fourth century BC) as having arisen from Pythagoras' interest in the practical side of commerce; Aristoxenus also considered him responsible for the creation of a system of weights and measures.

One of the first authors to write on the Pythagoreans was Aristotle, born in 384 BC; commenting on their relation with numbers he affirms [39] "since, then, all other things appeared to have been modeled on numbers in their nature, while numbers seemed to be the first things in the whole of nature, they supposed that the elements of numbers were the elements of all the things that exist, and that the whole heaven was harmony and number." Although Pythagoras and the Pythagoreans are nowadays believed to have contributed little to the techniques of mathematics [40], the importance of their contribution to the conceptual foundations of mathematics is undisputed. One of these contributions is the fuller realization of the abstract nature of numbers and geometric figures and the recognition that these are entities that belong to a class separate from that of physical objects.

Before the appearance of the first philosophers, Egyptians and Babylonians had a practical knowledge of mathematics, which was used in trade and in keeping the accounts of the temple and for the measurement of agricultural land. Although arithmetic operations with numbers were used, the meaning of the numbers was not discussed or examined in depth. The development of the number concept is related to the Pythagoreans; although numbers were used in reckoning from much earlier ages, the use of expressions such as "two oxen" did not imply knowledge of the full meaning of the word "two." In other words, mathematical concepts such as the number concept could only be understood when one abstracted something that was common in "two oxen" or "two people," and so on. The same process involves the geometrical concepts: a rectangular table and a rectangular plot of land have something in common that has to be identified as the rectangular shape; in this way, the concepts of rectangle, square, circle, etc. were born. Geometrical properties of simple figures were known, including the fact that in a right triangle the sum of the squares of the lengths of the sides adjacent to the right angle is equal to the square of the length of the opposite side, a theorem known nowadays as "Pythagoras' theorem."

The development of the number concept and the importance of numbers in the Pythagorean worldview have been associated with the introduction of coins, which first appeared in Lydia, in Anatolia, in the seventh century BC. One consequence of the emergence of a monetary economy, with the substitution of concrete goods by their abstract representation as numerical values in monetary units, may have been a stimulus for the Pythagorean discovery of the importance of numbers. The universalization of a system of weights and measures contributes in the same direction, helping to reveal the number as an abstraction. Measuring the weight of different goods with the same unit, for example, allows the same kind of association provided by currency: concrete goods of different natures may have the same weight, i.e., may correspond to the same number. In ancient Greece, many of the units were the same as those used in Egypt and in the east; the cubit as a unit of length and the talent, a unit of weight corresponding to about 58 pounds (25.8 kg) are examples. These units are already found in the verses of Homer and were therefore in use by the seventh century before our era.

Although the abstract nature of numbers was then recognized, numbers still retained a certain connection to physical objects; this appears, for example, in the representation of numbers by the Pythagoreans with pebbles or dots [41]. Among numbers of special interest were those that could be represented by a triangular array of dots, like 1, 3, 6, 10, and so on, called triangular numbers (the number 1, of course, was a "point-like" triangle). There were also the square numbers: 1, 4, 9, and so on. Some remains of this vision of the numbers have survived to our days, in the naming of the second power of a number as "square" of the number and the third power as "cube."

Pythagoras, or the Pythagoreans, established, probably using a monochord instrument, the *kanon*, that the musical intervals corresponded to some prescribed lengths of the string, which was varied by changing the position of a movable bridge. The pitch of the musical notes was inversely proportional to these lengths. Thus, the intervals of Greek music corresponded to the frequency ratios 1:2 (octave), 3:2 (fifth), and 4:3 (fourth), again confirming to them that the first four integers, 1, 2, 3, and 4 played a fundamental role in the world order [42].

A final comment is that it is difficult to separate myth from truth in the work and life of Pythagoras: the entry corresponding to the philosopher in the Oxford Companion to Philosophy states [43]: "Modern scepticism" about his achievements is "mostly justifiable."

Magnetism in Greece

There are but few references to magnetism in the writings of the early Greek philosophers; the comment on the magnet attributed to Thales, quoted at the beginning of this chapter, is the first ever recorded. Some philosophers, besides their considerations on other natural phenomena, not only described magnetism but also attempted to explain the cause that lay behind the bizarre behavior of the lodestone. A statement to this effect was reported by Aristotle in his treatise *De Anima* ("On the Soul"): "Thales, too, to judge from what is recorded about him, seems to have held the soul to be a motive force, since he said that the magnet has a soul in it because it moves the iron" [44].

In Diogenes Laertius' "Lives of the Philosophers" (early third century AD), the same idea is attributed to Thales:[15] "Aristotle and Hipias say that he ascribed souls to lifeless things too, taking the magnet and amber as his evidence." This idea attributed to Thales has been interpreted either as meaning that Thales shared the traditional beliefs of his contemporaries or that he had altogether abandoned them; the latter view is based on the fact that he had reserved this identification to spiritual beings only for admittedly out-of-the-ordinary objects, like the magnets [45].

Laertius lists amber side by side with the magnet as notable materials that are the object of Thales' remark. Amber is fossil tree resin, of yellow to orange color, transparent or milky, and is known for the property of attracting small objects when rubbed, due to the appearance of an electric charge (as referred

[15] J. Barnes, op. cit., p. 66.

in Chap. 1). Amber is also mentioned by Pliny the Elder; in his words,[16] "it attracts straw, dry leaves and bark from the linden-tree, just as a magnet attracts iron" (this property of amber will be discussed in Chap. 4).

In the Dialogues of Plato (427 BC–347 BC), one of the greatest Greek philosophers, one finds a description of how a magnet makes a piece of iron it touches behave as a magnet to another piece of iron, a phenomenon usually known as induction: "[46] (...) there is a divinity moving you, like that contained in the stone which Euripides calls a magnet, but which is commonly known as the stone of Heraclea. This stone not only attracts iron rings, but also imparts to them a similar power of attracting other rings; and sometimes you may see a number of pieces of iron and rings suspended from one another so as to form quite a long chain, and all of them derive their power of suspension from the original stone."

The Greek philosophers, after Thales, also tried to explain why the magnet had the power to attract iron. The philosopher Empedocles (c. 490–c. 435 BC) was one of them; he appealed to "effluences", or vapors, in his explanation. According to the book *Quaestiones*, written by a philosopher who was active around AD 200, Alexander of Aphrodisias, this was Empedocles' idea to explain magnetic attraction: "On the reason why the lodestone attracts iron. Empedocles says that the iron is attracted to the stone by the effluences which issue from both, and because the pores of the stone are commensurate with the effluences from the iron. The effluences from the stone stir and disperse the air lying upon and obstructing the pores of the iron, and when this is removed, the iron is drawn on by a concerted outflow. As the effluences from the iron travel towards the pores of the stone, because they are commensurate with them and fit into them, the iron itself follows and moves together with them."[17]

Along the same line, Democritus of Abdera (c. 460 BC–c. 370 BC), a Greek philosopher best known for his atomistic view of matter, also appeals to effluences to explain the properties of the magnet. In the words of Alexander of Aphrodisias: "Democritus also says that there are effluences and that like bodies are attracted to like, but adds that all are attracted to a void. Having made these hypotheses, he supposes that the lodestone and iron consist of similar atoms, but those of the stone are smaller and it is of rarer texture than the iron and contains more void. For this reason, its atoms being more mobile are attracted more quickly to the iron (for they are moving to their similar), and entering the pores of iron disturb the atoms in it as they pass between

[16] Pliny the Elder, Book XXXVII, no. 48, op. cit., p. 370.

[17] W. K. C. Guthrie, *A History of Greek Philosophy*, vol. II, quoting *Quaestiones*, p. 232.

owing to their small size. The atoms of the iron, thus disturbed, stream outside towards the stone because of their similarity and because it has more void. The iron [as a whole] follows them in their wholesale expulsion and movement and is itself drawn towards the stone. The reason why the stone does not move any more towards the iron is that the iron does not contain so much void."[18]

Epicurus of Samos (341 BC–270 BC), the philosopher known for his teachings that pleasure, or rather absence of pain, was the essence of life, had a different theory; according to the physician and philosopher Galen (c. AD 130–c. AD 200) [47], "His view is that the atoms which flow from the stone are related in shape to those flowing from the iron, and so they become easily interlocked with one another; (…)."

In every case, the proposed explanations involve mechanical actions, attraction by the void, interlocking of atoms, and so on (or maybe also animistic considerations, in the case of Thales with his remark on the "soul of the magnet"). Mechanical processes were certainly insufficient to explain magnetic effects; students of magnetism would have to wait for more than 20 centuries for a sound explanation of the phenomenon.

Chinese Records on Magnetism

Some of the earliest references to the lodestone come from the East, more specifically from China, where it was called 'tzhu shih'—the loving stone [48]. The first practical application of magnets—the compass—also originated in China, where it was used for navigation, possibly in the tenth century AD.

The first Chinese dynasty was the Shang dynasty, in the period from the eighteenth to the twelfth centuries BC, in the Chinese Bronze Age. Inscriptions found in pottery point to the origin of Chinese written language as early as 4000 BC. The unification of China occurred only in 221 BC, as the powerful feudal state of Ch'in established its dominance over the other states; in the process of unification, the written language was standardized.

During Chinese antiquity, there were enormous advances in the knowledge of natural phenomena and in different technologies. Astronomical knowledge reached a very high level, mostly stimulated by the need to perfect the calendar and also for divination purposes. Already by 1400 BC, the Chinese had determined the duration of the year as 365 1/4 days (the accepted value today is 365.242199 days), and the period of the lunar cycle of phases (called

[18] W. K. C. Guthrie, *A History of Greek Philosophy*, vol. II, quoting *Quaestiones* A165, p. 426.

synodic month, or lunation) as 29 1/2 days (known today to be 29.530588 days). They had a 354-day year and added, from time to time, an extra month with 29 or 30 days to keep the calendar in line with the motion of the Earth around the sun. They kept registers of astronomical phenomena that are invaluable today for the study of past astronomical events: they recorded eclipses since 720 BC, sunspots since 28 BC, comets since 613 BC, and novae and supernovae since 352 BC [49]. The collection and preservation of these records can in general be attributed to the activity of the State, and this strong presence of the State is a distinguishing mark of the early Chinese scientific development [50].

The Chinese started to make objects of cast iron as early as the sixth or fourth century BC, thanks to the invention of piston bellows that provided a steady flow of air into the smelting furnace. Among other technological breakthroughs, one may include paper, invented in China around AD 100; a mechanical clock in the eighth century; and gunpowder in the ninth century. In many technical fields, the Chinese led Europe until the sixteenth century.

The first important thinkers in ancient China were the Confucians and the Taoists, but there were also relevant contributions from members of other philosophical schools, known as Mohists, Legalists, and Logicians. The Confucians were the followers of K'ung-fu-tzu, or in Latinized form, Confucius, who lived from 552 BC to 479 BC. The Taoists (from Tao, "The Way") followed the teachings of Lao-tzu, said to have lived sometime between the sixth and fourth centuries BC, and the Mohists were followers of Mo-tzu (479 BC–381 BC). Although Confucians and Taoists were the most influential schools in Chinese intellectual history, and their teachings eventually evolved into two religions, Mohists, Logicians, and Legalists were more immediately related to the development of scientific ideas in China. The Mohists studied optics and mechanics and developed, together with the Logicians, the fundamentals of a scientific logic [51].They discussed the different forms of acquisition of knowledge: through hearsay, by inference, by direct observation, and by deliberate action, that is, by experimentation [52]. The Legalists, who flourished around 300 BC, preached in favor of laws that would strictly determine human conduct and came close to the concept of laws of nature. They also contributed to an incipient scientific attitude with their preoccupation with measure and quantification [53].

According to an anonymous manuscript at the time of the Han dynasty (206 or 202 BC to AD 220) (*The Nine Chapters on the Mathematical Art*), considered the most important work published in China [54] Chinese mathematical knowledge included the computation of areas of geometric figures, the volume of prisms, pyramids and cylinders, the solution of equations of the first degree, the solution of systems of linear equations, and the application of Pythagoras' theorem.

The Chinese studied the properties of magnets from a very early date and also developed the first applications of magnetism.[19] The first written record on the magnet is found in the *Lü Shih Chhun Chhiu* ("*Master Lü's Spring and Autumn Annals*"), written by a group of scholars and published in 240 BC, in the Chou period. It describes the property of the lodestone and asserts [55]: "The lodestone draws to itself iron particles." Other texts contain further observations on the lodestone: for instance, that it does not attract other metals or non-metallic objects [48]. As in this quotation from *Huai Nan Tzu* ("*The Book of the Princes of Huai Nan*"), written before 120 BC: "If you think that because the lodestone can attract iron you can also make it attract pieces of pottery, you will find yourself mistaken.(...) Fire is obtained from the sun by the burning-mirror, the lodestone attracts iron, crabs spoil lacquer, the mallow [malvaceous plant] turns its face to the sun. Such effects are very hard to understand." Also, from the same book: "Some effects are more pronounced at short range and others at long range. Rice grows in water but not in running water. The purple fungus grows on mountains, but not in stony valleys. The lodestone can attract iron but has no effect on copper. Such is the motion (of the Tao)" [48]. And from the *Lun Hêng* ("*Discourses Weighed in the Balance*"), a book from AD 83: "Amber picks up mustard-seeds and the lodestone attracts needles" [56].

In the third century BC, the Chinese used a diviner's board for fortune telling. It consisted of two parts, the upper one (circular) representing the heavens, resting on a square board describing the Earth, with divisions for the different compass points. The diviner threw small stones or figures onto the lower part and told the future from the positions they occupied. Among the figures used, one represented the constellation of the Great Bear (The Northern Dipper), in the shape of a spoon, made of wood, pottery, or stone. In the first century AD, (and possibly even before, in II BC) the spoon was made of lodestone, and from its behavior, it became known as the "south-pointing spoon" (Fig. 2.2). This was the first example of an application of the property of a magnet of orienting itself in the Earth's magnetic field. A remark on the property of this spoon is found in the *Lun Hêng* [57]: "But when the south-controlling spoon is thrown upon the ground, it comes to rest pointing at the south."

[19] This section relied heavily on J. Needham's monumental study *Science and Civilisation in China*, especially vol. 4, part I, Cambridge University Press, Cambridge, 1972.

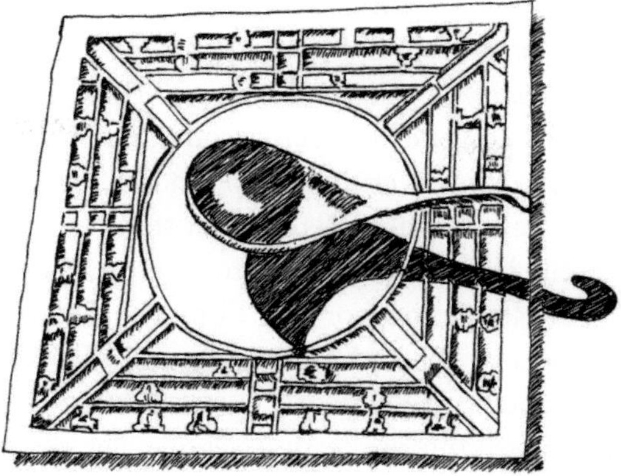

Fig. 2.2 Chinese divination board with "south-pointing spoon" made of magnetite

Chinese texts also mention a cart with a figure mounted on it that always pointed in the same direction. According to Needham, this had nothing to do with magnetism; it was a purely mechanical device [58].

In the first century AD, lodestones mounted on a pin, which allowed them to turn more freely, were common; in the seventh or eighth century, iron needles magnetized through contact with the lodestone substituted the pieces of rock, becoming the first needle compasses. The application of this early compass to navigation occurred perhaps in the tenth century, but the instrument was certainly being used by the eleventh century. The first Chinese description of a magnetic needle compass is from around AD 1080, one hundred years before the first European written records of such devices [59].

One of the first Europeans to discuss the use of the compass was the Englishman Alexander Neckam (1157–1217), in his book *De Nominibus Utensilium* ("On the Names of the Tools"), written around 1175–1183.

Roman Sources

The expansion of Rome in the third and second centuries before our era led to the Roman occupation of the Greek towns in the South of the Italian peninsula after the Pyrrhic War (280 BC–275 BC), and later, after the Macedonian Wars (214 BC–148 BC), assured complete control of Macedonia and mainland Greece. The Romans incorporated the achievements of Greek culture in the arts, in philosophy, and in science. However, no blossoming of creative ideas to match the Greek phenomenon is recorded.

A representative of the scientific tradition of the Roman world was the philosopher Lucretius (c. 98 BC–c. 55 BC), already mentioned. He described in his book *De Rerum Natura* ("On the Nature of Things") some properties of the magnet and how its power can be felt through a metallic obstacle: "I have seen Samothracian iron rings even jump up, and at the same time filings of iron rave within brass basins, when this Magnet stone had been placed under; such a strong desire the iron seems to have to fly from the stone" [60].

Another important Roman author was Pliny the Elder (AD 23–79), also quoted above. He was a writer and an investigator who left several treatises, the most famous of all being the *Natural History*. Pliny showed a deep interest in the magnet for its "striking properties": "For what phenomenon is more astonishing? Where has nature shown greater audacity?" [61] He describes the types of magnetic stones, distinguishing them as "male" or "female", according to the strength of their magnetic effects. He even speaks of a mythical stone from Ethiopia—*theamedes*—that "repels every kind of iron" [62]. He tells tales of two mountains near the river Indus, one that attracts iron and the other that repels it. Therefore, if a man has nails in his shoes, on one mountain "at each step he is unable to tear his foot away from the ground and on the other he cannot put his foot down" [63].

The only way one can observe repulsion with a magnet, of course, is by approaching another magnet, with the North pole near the North pole, for example. There appear to be no reports of interaction between magnets in the ancient texts, and consequently, no genuine observation of magnetic repulsion [64].

"There is a stone, colorless and without brilliance, (.) but it is preferred to all the most precious products of the Orient by those who know its virtues and its wonders." Thus wrote another Roman author, the fourth century poet Claudian, or Claudius Claudianus (c. 370–c. 404), in a text[20] that extolled the magic qualities of the magnet. And he continued: "Iron gives it life; iron recognizes and nourishes it; when the iron is withdrawn, it experiences the torments of hunger and thirst; it dies (…) Iron and the loadstone [sic] are drawn together and united. What can be this subtle flame which, entering these two metals, can give rise to this sympathy? What is the unknown charm which can unite them with a common will and a single desire?"

When Greek science entered a declining phase after the second century AD, its tradition was continued by the Arabs, whose scientific thought was strongly patterned after the Greeks [66]. Their scientific contribution will be discussed in Chap. 3.

[20] Claudian, "Idyl V," quoted by Alfred Still [65].

The next great step in the development of scientific explanations, after the Greeks had laid the groundwork, required a new critical attitude, a scientific revolution that enthroned the experimental method and the criticism of the classics as prime tools in the search for knowledge, in the sixteenth and seventeenth centuries. A revolution that brought with it a new view of the universe, with the Earth at the same time integrated into it and removed from its privileged position.

References

1. Moorey, P. R. S. (1994). *Ancient Mesopotamian materials and industries: The archaeological evidence*. Oxford University Press, reprinted Eisenbrauns, Winona Lake, Indiana, 1999, p. 84.
2. Campbell Thompson, R. (1936). *A dictionary of Assyrian chemistry and geology* (p. 85). Clarendon Press.
3. Heidelberger Akademie der Wissenschaften. (1981). *Akkadisches Handwörterbuch* (Vol. III, p. 1123). Otto Harrassowitz.
4. Reiner, E. (Ed.). (1989). *The Assyrian dictionary of the Oriental Institute of the University of Chicago* (Vol. 17, p. 36). Oriental Institute/J. J. Augustin Verlagsbuchhandlung.
5. Lucretius. (1978). On the nature of things. In *Great books* (Vol. 12(906), p. 92). Encyclopaedia Britannica.
6. Kramer, J. B. (1933–1934). The early history of magnetism. *Transactions of the Newcomen Society, 14*, 183–200, p. 185.
7. Kramer, J. B. (1933–1934). The early history of magnetism. *Transactions of the Newcomen Society, 14*, 183–200, p. 186.
8. Pliny the Elder. (1991). *Natural history: A selection*, Book XXVI, no. 127 (p. 359). Penguin Books.
9. Forbes, R. J. (1967). Extracting, smelting and alloying. In C. Singer, E. J. Holmyard, & A. R. Hall (Eds.), *A history of technology* (Vol. 1, p. 594). Oxford University Press.
10. Hodges, H. (1971). *Technology in the ancient world*. Penguin.
11. Johansson, L.-G. (2016). *Philosophy of science for scientists*. Springer.
12. Taton, R. (1966). *Histoire Générale des Sciences* (Vol. I, p. 204). Presses Universitaires de France.
13. Guthrie, W. K. C. (1967). *A history of Greek philosophy* (Vol. I, p. 33). Cambridge University Press.
14. Ronan, C. A. (1984). *The Cambridge illustrated history of the world's science* (p. 143). Cambridge University Press.
15. Acot, P. (1999). *L'Histoire des Sciences*, Collection Que Sais-Je? (p. 55). Presses Universitaires de France.

16. Pellegrin, P. (1996). Physique (p. 459). In J. Brunschwig, G. Lloyd, & P. Pellegrin (Eds.), *Le Savoir Grecque* (p. 463). Flammarion.
17. Lloyd, G. E. (1993). *Magic, reason and experience* (p. 243). Cambridge University Press.
18. Lloyd, G. E. (1993). *Magic, reason and experience* (p. 239). Cambridge University Press.
19. Aristotle. (1978). *Metaphysics, book I, 981b, great books* (Vol. 8, p. 500). Encyclopedia Britannica, Chicago.
20. Singh, S. (1999). *The code book* (pp. 217–242). Anchor Books, New York.
21. Altschuler, E. L., Calude, A. S., Meade, A., & Pagel, M. (2013). Linguistic evidence supports date for Homeric epics. *Bioessays, 35*(5), 417–420.
22. Michel, P. H., & Mugler, C. (1966). Les Sciences dans le Monde Gréco-Romain. In R. Taton (Ed.), *Histoire Générale des Sciences* (Vol. I, p. 208). Presses Universitaires de France.
23. Guthrie, W. K. C. (1967). *A history of Greek philosophy* (Vol. I, p. 371). Cambridge University Press.
24. Brumbaugh, R. S. (1981). *The philosophers of Greece* (p. 211). State University of New York Press.
25. Guthrie, W. K. C. (1975). *The Greek philosophers—From Thales to Aristotle* (p. 24). Methuen & Co.
26. Popper, K. R. (1975). *Objective knowledge* (p. 348). Oxford University Press.
27. Barnes, J. (1987). *Early Greek philosophy* (p. 25). Penguin.
28. Aristotle. (1987). Metaphysics 983b6-11, 17-27. In J. Barnes (Ed.), *Early Greek philosophy* (p. 63). Penguin.
29. Guthrie, W. K. C. (1967). *A history of Greek philosophy* (Vol. I, p. 59). Cambridge University Press.
30. Boyer, C. B., & Merzbach, U. C. (1989). *A history of mathematics* (2nd ed., p. 55). John Wiley.
31. Guthrie, W. K. C. (1967). *A history of Greek philosophy* (Vol. I, p. 46). Cambridge University Press.
32. Neugebauer, O. (1969). *The exact sciences in antiquity* (2nd ed., p. 142). Dover.
33. Dicks, D. R. (1970). *Early Greek astronomy* (p. 43). Chapman and Hall.
34. Barnes. (1987). *Early Greek philosophy* (p. 74). Penguin.
35. Barnes, J. (1987). *Early Greek Philosophy* (p. 79). Penguin.
36. Guthrie, W. K. C. (1967). *A history of Greek philosophy* (Vol. I, p. 151). Cambridge University Press.
37. Guthrie, W. K. C. (1967). *A history of Greek philosophy* (Vol. I, p. 181). Cambridge University Press.
38. Guthrie, W. K. C. (1967). *A history of Greek philosophy* (Vol. I, p. 213). Cambridge University Press.
39. Barnes, J. (1987). *Early Greek philosophy* (p. 208). Penguin.
40. Barnes, J. (1987). *Early Greek philosophy* (p. 210). Penguin.

41. Kline, M. (1972). *Mathematical thought from ancient to modern times*. Oxford University Press.
42. Guthrie, W. K. C. (1967). *A history of Greek philosophy* (Vol. I, p. 224). Cambridge University Press.
43. Honderich, T. (1995). Pythagoras. In *The Oxford companion to philosophy*. Oxford University Press.
44. Aristotle. (1952). *On the soul, "the works of Aristotle"* (Vol. I, p. 634). Encyclopaedia Britannica.
45. Guthrie, W. K. C. (1975). *The Greek philosophers—From Thales to Aristotle* (p. 66). Methuen & Co.
46. Plato, on Ion. (1952). *Great books* (Vol. 7, p. 144). Encyclopaedia Britannica.
47. Galen. (1952). On the natural faculties. In *Great books* (Vol. 10, p. 177). Encyclopaedia Britannica.
48. Needham, J. (1972). *Science and civilisation in China* (Vol. 4, part I, p. 232). Cambridge University Press.
49. Ronan, C. A. (1984). *The Cambridge illustrated history of the world's science* (p. 161). Cambridge University Press.
50. Ronan, C. A. (1984). *The Cambridge illustrated history of the world's science* (p. 185). Cambridge University Press.
51. Ronan, C. A. (1984). *The Cambridge illustrated history of the world's science* (p. 141). Cambridge University Press.
52. Haudricourt, A., & Needham, J. (1966). La Science Chinoise Antique. In R. Taton (Ed.), *Histoire Générale des Sciences* (Vol. I, p. 198). Presses Universitaires de France.
53. Ronan, C. A. (1984). *The Cambridge illustrated history of the world's science* (p. 142). Cambridge University Press.
54. Schwartz, R. (2008). A classic from China: The *nine chapters*. *The Right Angle, 16*(2), 8.
55. Needham, J. (1972). *Science and civilisation in China* (Vol. 4, part I, p. 31). Cambridge University Press.
56. Needham, J. (1972). *Science and civilisation in China* (Vol. 4, part I, p. 233). Cambridge University Press.
57. Needham, J. (1972). *Science and civilisation in China* (Vol. 4, part I, p. 262). Cambridge University Press.
58. Needham, J. (1974). *Physics and physical technology, Part II, Mechanical engineering* (Vol. 4, p. 245). Cambridge University Press.
59. Needham, J. (1972). *Science and civilisation in China* (Vol. 4, part I, p. 230). Cambridge University Press.
60. Lucretius. (1978). On the nature of things. In *Great books* (Vol. 12(1042), p. 94). Encyclopaedia Britannica.
61. Pliny the Elder. (1966). In M. R. Cohen & I. E. Drabkin (Eds.), *A source book in Greek science* (p. 311). Harvard University Press.

62. Cohen, M. R., & Drabkin, I. E. (1966). *A source book in Greek science* (p. 312). Harvard University Press.
63. Pliny the Elder. (1991). *Natural history: A selection*, Book II, no. 211 (p. 39), Penguin Books.
64. Wallace, R. (1996). "Amaze your friends!" Lucretius on magnets. *Greece & Rome, 178*, 185.
65. Still, A. (1946). *Soul of lodestone: The background of magnetical science* (p. 3). Murray Hill Books.
66. Arnaldez, R., Massignon, L., & Youschkevitch, A. P. (1966). La Science Arabe. In R. Taton (Ed.), *Histoire Générale des Sciences* (Vol. I, p. 445). Presses Universitaires de France.

Further Reading

Barnes, J. (1987). *Early Greek philosophy.* Penguin Books.
Cohen, M. R., & Drabkin, I. E. (1966). *A source book in Greek science.* Harvard University Press.
Guthrie, W. K. C. (1967). *A history of Greek philosophy* (Vol. I). Cambridge University Press.
Needham, J. (1972). *Science and civilisation in China* (Vol. 4, part I). Cambridge University Press.
Roberts, J. M. (1997). *The Penguin books history of the world* (3rd ed.). Penguin Books.
Ronan, C. A. (1984). *The Cambridge illustrated history of the world's science.* Cambridge University Press.

3

The Finger of God: Two Treatises on Magnetism

I guide the Pilot's course,
his helping hand I am,
The Mariner delights in me,
so doth the Merchant man.

The lodestone is the Stone,
the only stone alone,
Deserving praise above the rest,
whose virtues are unknown.

Robert Norman, The New Attractive *(1581) (Robert Norman, The New Attractive, quoted by R. Harré [1]).*

Summary The compass appeared in Europe about AD 1150. The first European work devoted to the magnetism of the lodestone, the "*Letter on the Magnet*," was published by the Frenchman Pierre de Maricourt (born c. 1220) in 1269. He wrote, "…this stone bears in itself the likeness of the heavens." However, the most important treatise on magnetism in this period was the "*De Magnete*" (1600) by William Gilbert (1544–1603), a physician at the court of Queen Elizabeth I (1533–1603). A significant work in the history of the sciences, it is also the first treatise on electricity. Gilbert's use of the experimental method and his critical attitude toward the classics were very innovative. The most baffling aspect of magnetism is how one object acts on another without a material link between them. In the sixteenth and seventeenth centuries, the idea of action at a distance gained increasing acceptance, culminating with the gravitational theory formulated by Isaac Newton (1642–1727). Newton considered that the Moon was attracted by the Earth in the same way

A. P. Guimarães, *A Longstanding Attraction*, https://doi.org/10.1007/978-3-032-02006-2_3

that our planet acts on any other object. Its motion deviated from a straight line, due to "falling" to the Earth.

Introduction

The Middle Ages began with the decline and subsequent fall of Rome to the invading Visigoth armies in the fifth century and lasted until the fall of Constantinople in the fifteenth century. During the Middle Ages, the relevant sources of scientific knowledge were Latin Europe, the Greek world, China, India, the Arab world, and pre-Columbian America. Only in Western Europe, however, would the scientific enterprise evolve into a new paradigm in the seventeenth century, a transformation generally described as the scientific revolution.

From the point of view of scientific development in the West, the Middle Ages can be divided, in a schematic way, into four different periods [2]: (1) the Dark age, or high Middle Ages (fifth–tenth centuries); (2) the awakening of Europe and the Islamic influences (eleventh–twelfth centuries); (3) the flourishing of the universities and the golden age of "scholastic" science (thirteenth century and beginning of the sixteenth century); and finally, (4) the interdependence of the sciences and technologies (1350–1450). The advances in the sciences and technology in the Middle Ages were very significant [3], comprising, in the first place, a change in the perspective of science, with the growth of the idea of control of humankind over nature and an attitude of disapproval toward restricting human speculation to the limits of one single scientific or philosophical system. The Middle Ages also witnessed the renaissance of the Greek idea of theoretical explanation in the sciences and the gradual shift in interest from metaphysical questions of causes to the mathematical description of phenomena.

In the physical sciences, notable progress stems from the first attempts (in the fourteenth century) to formulate a theory of motion. Among the important technological developments, there appear new machines devised to exploit waterpower and animal power, the mechanical clock and magnifying lenses, the astrolabe, and the quadrant. Also worth noting are the advances in medicine and surgery, in the descriptions of diseases, and in the descriptions of the fauna and flora of different regions.

A general trait to be found in the scientific thought of that period is the fact that questions posed by medieval scholars were frequently mingled with questions of the philosophy of science [4]; this remained true of science as practiced up to the seventeenth century and reflected the search for a new scientific paradigm.

Arab and Chinese Sources

The Arabs started the expansion of their empire out of the Arabian Peninsula in the seventh century. At its pinnacle, in the middle of the eighth century AD, the empire extended from Spain to India. The capital of the empire was Baghdad, founded in 762 by the caliph al-Mansur (709/714–775) of the Abbasid dynasty, on the site of an early Persian village. Baghdad reached a very high level of development in the succeeding years, when it reached half a million inhabitants, coming to be regarded as the richest city in the world in the period from the eighth to the ninth century.

In the seventh century, the Prophet Muhammad (c. 570–632) founded Islam as a new religion, with one omnipotent and omniscient God (Allah). The teachings of the Islamic sacred book, the *Qur'an*, (or Koran, which means in Arabic: reading or recitation), said to be revealed by God to Muhammad, valued medicine and knowledge in general, a factor that helped the development of the sciences in the Arab world. Arab scholars were particularly fruitful in their research in astronomy, producing some very precise tables that surveyed the positions of stars and planets. They developed mathematics, especially trigonometry and algebra, having created the word from the Arabic *al-jabr*, meaning "to restore", from the balancing of the two sides of an equation. They were responsible for the import, in the seventh century, of Hindu numerals, which were later universally adopted and are the symbols used today, known under the name of Arabic numerals. They were also the first to use the zero in the positional notation, as used today.

The original contributions of the Arabs are sometimes belittled in comparison with their important role in preserving and making available to Western Middle Age thinkers the works of the Greek philosophers and scientists. This effort to save the older manuscripts in fact started long before the flourishing of Arab sciences, when scholars took steps to preserve valuable texts before the destruction of the library and museum in Alexandria in the fifth century. Later, in important Arab centers of learning, especially in Baghdad and Cordoba, systematic programs of translation of the Greek works were carried out. Arab scholars not only translated the classic texts but also made many additions and commentaries that very much enhanced their value. Under the Arabs, science had for the first time an international character, thanks to the extension of the kingdom in the mid-eighth century, from the Iberian Peninsula to the Indian subcontinent, and favored by the unifying forces of common religion and language [5].

The greatest contribution to physics by Arab scholars was the work of ibn al-Haytham (known in the West as Alhazen) (c. 965–1039), born in Basra, Iraq. He worked in the library in Cairo during the reign of the caliph al-Hakin. He wrote a treatise, called "Treasury of Optics" (*Kitab al-manazer*), and made important steps toward the understanding of optical phenomena. He rejected the traditional Greek picture of light emitted by the eyes, used the concept of "light rays" that propagated in straight lines, and correctly interpreted that refraction of light in the boundary between two transparent media was due to the difference in velocity of propagation in the two media [6].

References to magnetism are not very common among Arab authors; the first text on the magnetic compass by an Arab author (writing in Persian) contained the stories compiled by Muhammad al-Awfa (d. c. 1230), in the *Jami al-Hikayat*, published around 1232. In 1282, Bailak al-Qabajaqa of Cairo was the first to write in Arabic on the use of the magnetic compass for navigation.[1] In the treatise of Tayfashi (thirteenth century), the existence of the two poles of the magnet is reported, and also its property of indicating direction.[2] Arab mariners may have used the compass as early as in the eleventh century.[3]

In China, the existence of magnetic poles and the fact that the magnetic North did not coincide with the geographical North were known by the eighth or ninth century. The latter discrepancy would be discovered in the West only seven centuries later. This difference in direction between the geographical North–South line (the meridian) and that of the compass is called magnetic declination. It arises from the fact that, although the compass needle points roughly in the North–South direction, the Earth's North and South magnetic poles are not precisely located on the corresponding geographical poles, which are the points that mark on the Earth's surface the axis of rotation of the planet. The position of the magnetic poles on the surface of the planet has varied over geological time and in fact, North and South magnetic poles have reversed several times, on average about once every 280,000 years in the last six million years [8]. This effect is not completely understood; we will return to the subject of terrestrial magnetism in Chap. 7.

---

[1] See also Schmidl [7].

[2] R. Arnaldez et al., p. 512.

[3] R. Arnaldez et al., p. 498.

Magnetism and the Compass in Europe

The diffusion of scientific knowledge in Europe during the early Middle Ages owed much to works of encyclopedic character, such as Pliny's *Natural History*, the *Etymologies* of Isidore of Seville (560–636), and the *Geometry of Boethius*, by Anicius Manlius Severinus Boethius (c. 480–c. 525). These works, although very valuable, are mostly compilations of earlier knowledge. They were followed by the works of Venerable Bede (673–735), Alcuin of York (735–804), and Hrabanus Maurus (776–856), each one transcribing much from their predecessors. At the end of the tenth century, a noteworthy scholar in the Christian West was Gerbert of Aurillac (c. 945–1003), who wrote the logical treatise *De rationali et de ratione uti* ("Concerning the Rational and the Use of Reason") and became the first French pope, as Sylvester II. He had contacted Arab scholars during a stay in Spain, and after that helped to propagate the use of the Arabic numerals.

After the sixth century, when St. Benedict of Nursia (or Norcia) (c. 480–c. 547) founded the monastery of Monte Cassino, in Italy, the preservation and diffusion of scientific knowledge took place in the monasteries and in the adjoining schools. The monks of the order, known as the Benedictine, maintained large libraries and systematically copied religious, philosophical, and some literary works. In the eighth and ninth centuries, schools were created at the cathedrals, for example, in York in England and Orléans in France. They were initially dedicated to the education of priests, but later accepted lay students. Starting from the ninth century, when the University of Bologna was created, until the fifteenth century, a movement of creation of centers of higher learning swept over Europe: they were founded in Oxford (1167), Paris (1170), Salamanca (1218), Padua (1222), Cambridge (1290), Rome (1303), Florence (1321), Vienna (1365), Heidelberg (1386), Saint Andrews (1411), and Uppsala (1477), among others. These new centers were to become important nuclei of scholarship in the sciences.

Many Western authors of the early Middle Ages described the properties, real or legendary, of the magnet. Among these, one can name the Latin grammarian Caius Julius Solinus (second and third century AD), the theologians Augustine (354–430) and Isidore of Seville [9] already mentioned. Later, the Bishop of Rennes, Marbode (1035–1122), in the *Liber de Gemmarum* ("Book on the Gems"), helped to propagate many legends related to the uses of the lodestone.

While the application of magnetism to the construction of the magnetic needle compass was described in China in a publication around AD 1080, in the West, the first references to the compass appeared about one century later. The available evidence leads one to think that the compass evolved independently in the East and the West.[4]

Three authors are nowadays regarded as pioneers in reporting the use of the magnetic compass in the West: these are the Englishman Alexander Neckam (1157–1217), the Frenchman Guyot de Provins (fl. 1184–1210), and Jacques de Vitry (c. 1165–1240), in this order of chronological priority.[5]

Alexander Neckam was born in St. Albans, some 20 miles (32 km) northwest of London, in 1157, and lectured in Paris and Oxford; he was appointed abbot of Cirencester in 1213. Neckam is the author of several books, among them *De Nominibus Utensilium* ("On the Names of the Instruments") and *De Naturis Rerum* ("On the Nature of Things"), both containing mentions of the magnetic compass. The earliest reference is to be found in *De Nominibus Utensilium*, written around 1175–1183:[6] in a ship, there should be "a needle mounted on a pivot, which will oscillate until the point looks to the east [*sic*], and the sailors will know how to direct their course when the northern constellation of the Little Bear is obscured by the troubled [state of the] atmosphere; for it never disappears below the horizon, because of its small circle." A second mention by Neckam is given in his book *De Naturis Rerum*, of 1197–1204: "So sailors crossing the sea, when because of overcast skies they lose the sun's light, or when the world is wrapped in the darkness of night, and they do not know what cardinal point the ship is headed toward, put a needle above the lodestone; and the needle revolves until, after its motion has stopped, its point faces due north" [11].

Neckam recounts the old legend of the iron statue of Muhammad suspended in mid-air through the action of magnets; the same tales of magnets sustaining statues also appear in Pliny's *Historia Naturalis*. A similar legend tells that in India, the Sun temple of Konarak, in the state of Orissa, had a statue held from a large lodestone mounted under the vault.[7]

Guyot de Provins (end of XIIth century - beginning of XIIIth), a cleric and minstrel at the court of the Emperor Frederick I, or Barbarossa, wrote *La Bible*, a satirical poem published probably in 1206, where he makes a

[4] J. A. Smith [9]. This is the basic reference used here on the work of Peregrinus and his predecessors.

[5] This is found in the article of J. A. Smith, who gives arguments against dates estimated by other science historians.

[6] Alexander Neckam, "De Nominibus Utensilium," quoted by J. A. Smith [10].

[7] Debala Mitra, *Konarak*, published by the Director General, Archaeological Survey of India, New Delhi, 1976, p. 10, quoted by C. K. Majumdar [12].

metaphor involving the magnetic compass and proceeds to describe its use by the mariners. The other claimant to priority is the historian and cleric Jacques de Vitry (c. 1160/70-1240), who refers to the compass in his book *Historia Orientalis Hierosolymitana* ("Eastern History of Jerusalem"), a text on which he worked for many years; the part on the compass seems to have been written in 1204.

These references treat the magnetic compass as an instrument already known for some time; from this fact, one is led to date tentatively the appearance of the compass in Europe at about AD 1150 [13]. Compasses mounted on pivots ("dry" mounting) and compasses floating on water ("wet" mounting) are equally present in these early references, which suggests that these two forms of compasses appeared in Europe at approximately the same time.

Early medieval authors often used the word adamant, from the Greek *adamas*, "invincible," meaning either the lodestone or a very hard material, sometimes the diamond, and in some other passages, steel. The source of confusion between these meanings is the incorrect attribution of the origin of the word to the Latin *adamare*, or "having an attraction for." Therefore, *lapide adamanten* became the attracting stone, from which derived the Romance languages *aimant* (French), *imán* (Spanish) and *ímã* (Portuguese) [14].

The French philosopher and theologian Guillaume d'Auvergne (also referred to as William of Auvergne, or William of Paris) (c. 1180–1249), a follower of Augustine, wrote *Magisterium Divinale* ("The Divine Teaching"), his major work, in the period 1223–1240. Guillaume d'Auvergne was a professor in Paris, appointed around 1225, and became the bishop of Paris in 1228. In his work *De Universo Creaturarem* ("On the Universe of Created Things"), he discusses the way a magnet attracts one piece of iron, and then another piece in contact with the first piece, the phenomenon of "induction" (Chap. 2). He is one of the first authors to suggest that magnetic forces are active also in the sky; he uses the idea of induction to explain how the motion is transferred, at a distance, from one heavenly sphere to the other.

The Belgian physician John of St. Amand (fl. 1261–1298) was a canon of Tournai, in the Southwest of Belgium; among the many books he wrote, the *Expositio sive Additio Super Antidotarium Nicolai* ("A Commentary on the Antidotary of Nicholas of Salerno") contains passages that anticipate the idea that the Earth is a magnet, an idea that was developed later by the Englishman William Gilbert, at the end of the sixteenth century. He states [15, 16], "Wherefore I say that in the magnet is a trace of the world, wherefore there is in it one part having in itself the property of the west, another of the east, another of the south, another of the north."

Magnetic Epistle from the Trenches

Charles of Anjou (Charles I) (1226–1285) and his brother, the French King Louis IX, participated in the ill-fated Crusade in Egypt in the years 1248–1250. In the 1260s, Charles led a military campaign in Sicily and in mainland Italy, conquering the island and Naples, defeating King Manfred (c. 1232–1266??), of the German Hohenstaufen dynasty, then in power. In 1269, he held under siege and took Lucera, in Apulia (Puglia), near the city of Foggia, in the heel of the Italian boot. In the trenches encircling Lucera, in that year, the first European work entirely devoted to the discussion of the magnetic properties of the lodestone, the *Epistola de Magnete* ("Letter on the Magnet"), was written (Fig. 3.1).[8] It was the work of the Frenchman Pierre de Maricourt (b. c. 1220), also known by his Latin name, Petrus Peregrinus [17]. He may have received the title of "Peregrinus" not for visiting the Holy Land or participating in the Crusades, as was usual, but for his role in that military campaign. Pope Clement IV (d. 1268), also a Frenchman, had declared the attacks against the Hohenstaufens as official Crusades since the Germans had the support of the Saracens, the denomination under which the followers of Islam were known at the time.

The work of Peregrinus was written in the form of a letter to a soldier, a certain Siger, of Foucaucourt, in Picardy, France. In this short letter, written in response to questions put by the same Siger, Pierre de Maricourt relates some properties of the magnet, describes experiments with it, and mentions its applications. The influence of the *Epistola* was very large: it was cited by many books, such as William Gilbert's *De Magnete* ("On the Magnet") (1600) and Athanasius Kircher's (c. 1602–1680) *Magnes, sives De Arte Magnetica* ("The Magnet, or About the Magnetic Art") (1641), and plagiarized in *De Natura Magnetis* ("On the Nature of the Magnet") (1562) of Jean Taisnier [18]. There are at least 31 known manuscript versions of the *Epistola*.

Little is known about Pierre de Maricourt. At the time of the letter, he was most likely an engineer in the army; he probably came from the town of Méharicourt, in Picardy, a province corresponding roughly to the present French *région* of the same name, in the North of the country. He seemed to be primarily interested in designing and building scientific instruments; he declares the *Epistola* "part of a work on the construction of philosophical instruments".[9]

[8] It is not known when the letter was written, only that it was concluded in 1269.

[9] Pierre de Maricourt (a), p. 369.

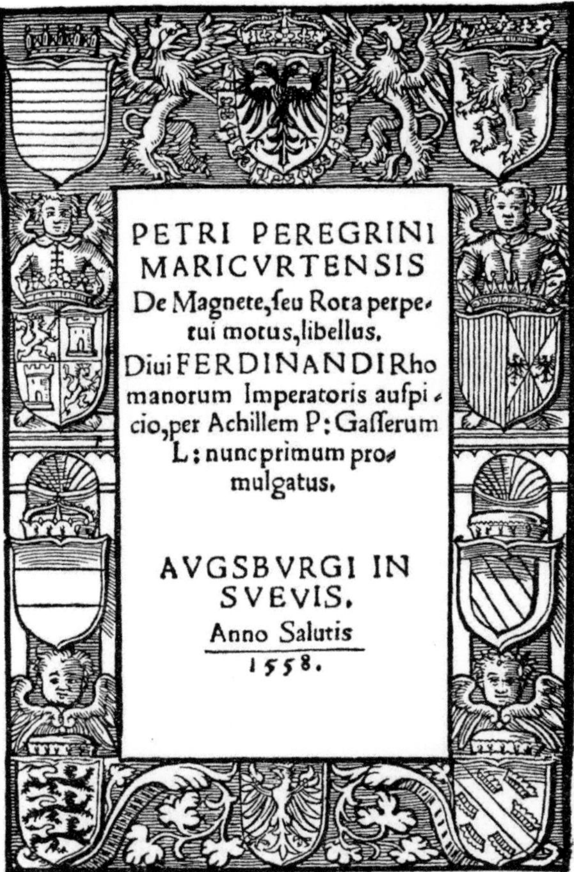
PETRI PEREGRINI
MARICVRTENSIS
De Magnete, ſeu Rota perpe-
tui motus, libellus.
Diui FERDINANDI Rho
manorum Imperatoris auſpi-
cio, per Achillem P: Gaſſerum
L: nunc primum pro-
mulgatus.

AVGSBVRGI IN
SVEVIS.
Anno Salutis
1558.

Fig. 3.1 Title page of the *Epistola de IMagnete* (Letter on the magnet) (1269), by the Frenchman Peter Peregrinus (b. c. 1220), in an edition of 1558. (Courtesy Bakken Museum)

"The disclosing of the hidden properties of this stone is like the art of the sculptor by which he brings figures and seals into existence," Pierre de Maricourt writes in the first part of the work. He then continues to list the physical properties that characterize the specimens of lodestone: the color, homogeneity, weight, and strength; by strength, he means the power to attract pieces of iron.

He describes the two regions of the lodestone, which reveal a stronger magnetic attraction, as "poles" and chose this denomination in analogy to the celestial sphere: "I wish to inform you that this stone bears in itself the likeness of the heavens, as I will now clearly demonstrate. There are in the heavens two points more important than all others, because on them, as on pivots, the celestial sphere revolves: these points are called, one the arctic or north pole, the other the antarctic or south pole." He refers to the celestial poles, not to the pole star, since he knew that the pole star does not sit exactly on the celestial North Pole. Peregrinus' association of the properties of the lodestone to the heavens was in line with scholastic philosophy, or more specifically with the thought of his contemporary St. Thomas Aquinas (1224/25–1274), the greatest philosopher of the Christian Middle Ages, who also attributed the effects of the magnet to *virtus coeli*, the "power of heavens".[10]

Peregrinus discusses how a lodestone, broken into two pieces, shows two pairs of poles, how a piece of iron becomes magnetic after touching a lodestone, and how like poles are repelled and opposite poles attracted. Although earlier authors had already described magnetic repulsion, the association of repulsion with equal poles is made here for the first time.

The property of the lodestone of aligning in the North–South direction was already known; this is a very remarkable property, of course, impossible to predict with the knowledge of natural phenomena then available. Why would the stones that attracted pieces of iron have anything to do with the cardinal points on the Earth's surface? When one considers closely this extraordinary fact, one understands why the Irish physicist John Desmond Bernal (1901–1971) regarded this as the greatest discovery in the history of physics! [20] Equipped with the knowledge of this property, Peregrinus then proceeds to show how to make a floating compass with a scale of 360°: this instrument is regarded as the first compass with the proper scale division.[11] He also described, besides this "wet" compass, a "dry" one, i.e., one where the needle turns on pivots (Fig. 1.1).

Peregrinus was the first to shape a lodestone; he cut it as a sphere but stopped short of making the analogy of the spherical magnet with the planet Earth. Since the magnet seemed to be linked to the heavens, he believed (erroneously) that a spherical lodestone pointed to the pole, if free to turn, would rotate continuously, following the motion of the stars in the sky. Such a lodestone would then make a complete turn in 24 h and could be used as a clock.

He rejected the idea that the lodestone tended to point to the poles due to the presence of magnetic rocks at the Earth's poles. He argued that if this were true, since there are iron mines in many parts of the world, the compass would point to completely different directions at different places.

In the last part of his *Epistola*, Peregrinus gives the design for a perpetual motion machine based on magnetic attraction and repulsion. To justify his

[10] Aquinas, St. Thomas. (1612) "*Opera Omnia,*" Antwerp, vol. 8, *Quaestio Unica: de Anima, art. 1 (Utrum anima humana possit esse forma et hoc aliquid)*, p. 437, quoted by King [19].

[11] Pierre de Maricourt (a); note on page 374.

proposal, he affirms:[12] "I have seen many persons vainly busy themselves and even become exhausted with much labor in their endeavors to invent such a wheel. But these invariably failed to notice that by means of the virtue or power of the lodestone all difficulty can be overcome." In his project, he had the teeth of an iron cogwheel attracted by a lodestone; as the lodestone approached each tooth, it turned an axle. This model of a perpetual motion machine, although condemned to failure like all the others, may have been the first attempt to design such a machine based on scientific premises.[13]

In his *Opus Tertium* ("Third Work"), written in 1267, the English philosopher Roger Bacon (1214–1294) refers with words of praise to an investigator who is "a master of experiments (*dominus experimentorum*)." "I know of only one person who deserves praise in the works of this science," he adds in another passage, speaking of the use of burning mirrors.[14]

The main importance of the *Epistola* resides in the fact that it provided the first published systematization of the phenomena of magnetism and that its contents were, in general, based on experimental evidence; however, it contained very little in terms of new knowledge [9].

By the fourteenth century, the use of the compass needle was widespread, and the phenomenon of magnetic attraction had already established itself as a powerful image. This is exemplified by the allure of the voice of St. Giovanni Bonaventura in the verse of Canto XII, Paradise, in the *Divine Comedy*, written by the great Italian poet Dante Alighieri (1265–1321) in the first decades of the century:[15]

Out of the heart of one of the new lights
There came a voice, that needle to the star
Made me appear in turning thitherward.

In the fifteenth century, Nicholas of Cusa (Nicolaus Cusaneus) (1401–1464) in *Exercitationes* likens Christ to the compass, both of them showing the way to men and both deriving their power from the heavens [11].

[12] Pierre de Maricourt (a), p. 375.

[13] Pierre de Maricourt (a), p. 375.

[14] Quoted by E. Grant (a), see Ref 20, p. 533.

[15] Translation by Henry Wadsworth Longfellow. In the Italian original: "del cor de l'una de le luci nove si mosse voce, che l'ago a la stella; parer mi fece in volgermi al suo dove"

Wonderful Is the Lodestone

In 1558, Giovanni Battista (or Giambattista) della Porta (1535–1615), a Neapolitan researcher and playwright, published the first edition of *Magia Naturalis* ("*Natural Magic*"), a work containing a mixture of natural facts, astrology, and practical knowledge. The second expanded version appeared in 1589, incorporating more scientific facts, including experiments with the lodestone. Della Porta was the first to propose a way of measuring the strength of the magnet using a balance. The force exerted on a piece of iron was compensated for by a weight put on another pan of the balance, and this weight was the measure of the lodestone's attraction. Della Porta was also a precursor in founding the first association for the study of scientific matters, the *Academia Secretorum Naturae* ("Academy of the Secrets of Nature"); their members called themselves *Otiosi* ("Men of Leisure"). It was followed by the foundation in Rome of the Accademia dei Lincei, the first true scientific society, in 1603; Galileo Galilei was one of its members.

Near the end of the sixteenth century, Robert Norman (fl. 1590), an English instrument maker, published in London *The New Attractive* in 1581; the book, on magnetism, was reprinted three times before the end of the century. Norman had been a sailor for 20 years and was well acquainted with the practical use of the compass. He wrote of the magnetic compass:[16]

I guide the Pilot's course,
his helping hand I am,
The Mariner delights in me,
so doth the Merchant man.

The lodestone is the Stone,
the only stone alone,
Deserving praise above the rest,
whose virtues are unknown.

He promises in the preface to "ground his arguments only upon experience, reasons and demonstrations".[17] In this book, he described the magnetic dip, or inclination of the compass, in relation to the horizontal, which he had discovered in 1576. In his own words, "(...a newe discouered secrete and subtill propertie, concernying the declining of the needle, touched therewith

[16] Robert Norman, *The New Attractive*, quoted by R. Harré [1].

[17] Robert Norman, *The New Attractive*, London, 1581, quoted by E. Zilsel [21].

under the plaine of the horizon)".[18] This effect of inclination in relation to the horizontal plane had already been found before by the German instrument maker Georg Hartmann (1489–1564), in 1544.

Norman performed many experiments with compasses, and his discoveries afforded him "incredible delight." He thought that all compasses placed on different points of the Earth's surface would point to the same "point respective." This conviction represents an important shift from Peregrinus' "power of the heavens," relating for the first time the influence upon the magnetic needle to the planet Earth.

The most important treatise on magnetism in this period, and for many years to come, was the book *De Magnete* ("On the Magnet") (Fig. 3.2), published in London in 1600 by William Gilbert (1544–1603) (Fig. 3.3), physician at the court of Queen Elizabeth I. Elizabeth I reigned in England in the second half of the sixteenth century, in a period characterized by a flourishing of the arts and literature. During the Elizabethan period, England also reached a zenith in its economic power and global influence.

Elizabethan London witnessed, at about the same time that *De Magnete* was published, the first presentation of Hamlet (1600 or 1601), one of the greatest creations of the playwright and poet William Shakespeare (1564–1616). Hamlet tells the story of a Danish prince, based on a legendary figure, who is drawn into a series of tragedies as he tries to avenge the murder of his father. The words of one of Shakespeare's characters, John of Gaunt, Duke of Lancaster, uncle of Richard II, are appropriate to describe England in this period of glory:[19]

This royal throne of kings, this scepter'd isle
This earth of majesty, this seat of Mars
This other Eden, demi-paradise

William Gilbert was born in 1544 in Colchester (Essex), northeast of London. He studied in Cambridge, where he obtained his A. B. (Bachelor of Arts) in 1561 and graduated in medicine in 1569. He was a talented experimentalist and dedicated himself to studying electrical and magnetic phenomena; he created the word "electric" from *elektron*, the Greek for amber. He was appointed president of the Royal College of Physicians and became the personal physician of Queen Elizabeth I in 1600. Besides the book *De Magnete*, published in 1600, he wrote another book on cosmology, *De Mundo* (1651),

[18] Robert Norman, *The New Attractive*, London, 1581, quoted by W. J. King [22].

[19] W. Shakespeare, 'Richard II', Act 2, Scene 1.

GVILIELMI GIL-
BERTI COLCESTREN-
SIS, MEDICI LONDI-
NENSIS,

DE MAGNETE, MAGNETI-
CISQVE CORPORIBVS, ET DE MAG-
no magnete tellure; Phyſiologia noua,
plurimis & argumentis, & expe-
rimentis demonſtrata.

LONDINI
EXCVDEBAT PETRVS SHORT ANNO
MDC.

Fig. 3.2 Title page of *De Magnete* (on the magnet) (1600), by William Gilbert (1544–1603), the first treatise on magnetism. (Courtesy Bakken Museum)

published posthumously in Amsterdam by his half-brother, who had collected his scientific papers.

His most important book is *De Magnete*,[20] one of the more relevant works in the history of the sciences. It represents an important milestone in the change of the attitude toward nature and the sciences, which occurred in the sixteenth and seventeenth centuries. Gilbert created a complete treatise of

[20] The complete title is *De Magnete Magneticisque Corporibus et de Magno Magnete Tellure Physiologia Nova* ("On the Magnet and Magnetic Bodies, and on The New Physiology of That Great Magnet the Earth)."

Fig. 3.3 William Gilbert (1544–1603), English physician and experimentalist, author of *De Magnete* (1600)

magnetism, but his aims were higher than that: he expected to inaugurate a new cosmology in which magnetism played a central role, or a new philosophy of nature, or physiology, as he called it.

The work was written in Latin and was divided into six books, each of which was divided into several chapters. In the Preface, he announces, to leave no doubts as to his allegiance to experimental science, that the work was dedicated to those "who not in books but in things themselves look for knowledge" [23]. In the first chapters of Book I, he makes a critical revision of the writings on magnetism, condemning the myths that passed unchecked from one author to the next, tales that would not survive empirical tests, "figments and falsehoods" "dealt out to mankind to be swallowed".[21]

He conducted many experiments with a spherically machined lodestone; since it represented a model of the Earth, he called it "terrella", or small Earth, an early example of an experimental scale model (Fig. 3.4). He described how the poles of the *terrella* could be determined, following the same methods found in the text of Peregrinus. He experimented with the lodestone, showing that a split magnet produced two new magnets; he proved that the lodestone did not attract other metals than iron and did not affect wood, glass, and bone.

[21] W. Gilbert (a), op. cit., p. 3.

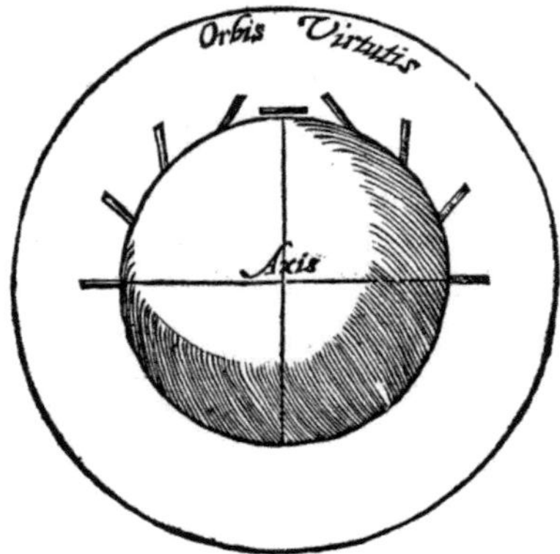

Fig. 3.4 Drawing of a *terrella* with compasses on different points of its surface, showing the effect of inclination (or "dip"), from De *Magnete* (1600). (Courtesy Bakken Museum)

Gilbert compared the alignment of the compass in the North–South direction with the fact that the Earth's axis of rotation had a constant direction in space (neglecting other movements, such as precession and nutation): "Like the earth, the loadestone [sic] has the power of direction and of standing still at north and south; it has also a circular motion to the earth's position, whereby it adjusts itself to the earth's law".[22]

For him, there exists a link between the Earth's magnetism and the magnetism of the lodestone extracted from the mines: "Thus every separate fragment of the earth exhibits in indubitable experiments the whole impetus of magnetic matter; in its various movements it follows the terrestrial globe and the common principle of motion".[23]

In Book II, he discusses several phenomena connected with magnetism under the general denomination of "movements." These are: (1) Attraction, which he terms coition (from the Latin *co* + *ire*, going together), (2) The alignment with the North–South direction, (3) Declination, or deviation from the meridian, which he characterizes as a "perverted" motion, (4) Magnetic dip, or inclination below the horizontal plane, and (5) Revolution, or circular motion.[24]

In Book II, he describes electric phenomena, making the *De Magnete* the first published treatise on electricity. In his study, he starts with the properties of amber; all the materials that attract chaff when rubbed he calls "electrics";

[22] W. Gilbert (a), op. cit., p. 24.

[23] W. Gilbert (a), op. cit., p. 25.

[24] W. Gilbert (a), op. cit., p. 26.

the others were "non-electrics." According to Duane H. Roller, author of a study on *De Magnete*, with the establishment of this division, Gilbert separated magnetic and electric phenomena, which existed side by side as effects related to "attraction." With this separation, he founded the science of electricity [24].

In *De Magnete*, the electric attraction is demonstrated experimentally through its effect on a needle (not a magnetic one) mounted on a pivot, a simple instrument known as a *versorium*. Gilbert observed that this attraction of electric origin is reduced in cloudy weather or when the object is exposed to the "moisture from the mouth." What he had thus proved was, putting it in modern terms, that moisture reduced the insulation of the object, allowing the electric charges to flow away and therefore discharging it.

Gilbert tended to believe that the explanation for the electric attraction is due to some sort of material emission from the amber: "Hence it is probable that amber exhales something peculiar that attracts the bodies themselves, and not the air".[25] He proves that the intervening air is not involved by demonstrating that the flame of a candle is not affected by the proximity with the attracting amber.[26]

He also proves experimentally that heat momentarily destroys the magnetic properties of objects otherwise attracted by a magnet: a magnetized needle stands still near a piece of red-hot iron. He finds that the same piece of iron has its magnetic properties restored as it cools down.

He divides a lodestone into two parts and finds that "If magnetic bodies be divided or in any way broken up, each several part hath a north end and a south end";[27] rejoining two halves of a round lodestone cut along a parallel re-establishes the original magnet, with the poles in the original places.

Gilbert speaks of the property of the space around a lodestone: in modern terms, we would call it the magnetic field. He called this sphere of magnetic influence the *orbis virtutis*. He finds that inserting either boards, pottery, or marble between the magnet and a piece of iron does not interfere with the attraction; only a plate of iron can affect the influence of the magnetic force. In his studies, he also notes that the magnetic effect of the lodestone is reduced as the distance is increased.

Gilbert discusses at length one of the most important properties of the magnetized bodies—their tendency to align in the North–South direction: "A little iron bar—that soul of the mariner's compass, that wonderful director in

[25] W. Gilbert (a), op. cit., p. 31.
[26] W. Gilbert (a), op. cit., p. 31.
[27] W. Gilbert (a), op. cit., p. 53.

sea-voyages, that finger of God, so to speak—points the way and has made known the whole circle of earth, unknown for so many ages"[28] And more: "(...) no invention of human arts has ever been of greater use to mankind".[29]

Gilbert's experiments demonstrated that a floating magnetic needle is not attracted as a whole to the Pole, repeating the experiments and conclusions of Robert Norman. In Gilbert's words[30] the "direction is not produced by the attraction, but by a disposing and conversory power existing in the earth as a whole. (In modern terms, that fact might be explained as an effect arising from the action of two equal forces applied at different points (the poles of the magnetic needle) that turn the needle but do not move the center of mass of the compass.)

He considered several circumstances under which iron is naturally magnetized due to the influence of the Earth. For example, by allowing a piece of white–hot iron aligned in the North–South direction to cool or by hammering or stretching an iron bar in this same direction. The same phenomenon is observed with a piece of iron that is part of the structure of a building, kept pointing along the North–South direction for many years.

In *De Magnete*, Gilbert studied the phenomenon of declination, or the deviation of the compass from the true North–South direction, by experimenting with the *terrella*. He proved that in the *terrella*, deviations occur when there are large irregularities on its surface; he argued that the same idea must then be applicable to the Earth. Again, using the *terrella*, he demonstrates that the compass needle is parallel to the surface of the lodestone at a point near the magnetic equator and is perpendicular to the surface at the poles. Therefore, this deviation from the plane parallel to the surface (or horizontal direction)—the dip—is beautifully explained. The angle formed by the needle with the horizontal, at intermediate latitudes, has a value between 0° and 90° (Fig. 3.4).

In another experiment, he moved the magnetic needle away from the *terrella* and recorded the directions of the needle at several points. The figure representing the directions chosen by the needle located at circles of different radii is a rough anticipation of the images of the magnetic lines of field obtained by Michael Faraday (1791–1867) using iron filings, more than 200 years later. In a remarkable comment on the abstract nature of this concept, we would describe today as field lines, he states, "(.) and so the spheres are magnetical, and yet are not real spheres existing by themselves".[31]

[28] W. Gilbert (a), op. cit., p. 75.

[29] W. Gilbert, quoted by Duane Roller.

[30] W. Gilbert (a), op. cit., p. 82.

[31] W. Gilbert (a), op. cit., p. 103.

The rotation of the Earth is related in many ways to magnetism, according to Gilbert; he deals with this question at length in Book VI. He also discusses the application of magnetism to different questions, his "magnetic philosophy," or cosmology. "The causes of the diurnal motion are to be found in the magnetic energy and in the alliance of bodies",[32] he explains, "(produced partly by the energy of the magnetic property and partly by the superiority of the sun and his light)".[33]

He rejects the Aristotelian belief that the Earth remained immobile at the center of the universe. He argues that the apparent turning of the stars is due to the rotation of the Earth: "It is more accordant to reason that the one small body, the earth, should make a daily revolution than that the whole universe should be hurled around it".[34] Without this rotation, one side of the earth would freeze with intense cold, and the other would be scorched, making life on the planet impossible.[35] He also recognizes the importance of the angle between the axis of the rotary motion of the Earth and the line perpendicular to the plane that contains the planet and the sun; without this angle, there would be no seasons.[36] He is also aware that the axis of the Earth does not in fact point along a fixed direction but turns in a slow motion called the precession of the equinoxes.[37]

In Book IV Gilbert discusses declination, which he attributes to irregularities in the surface of the Earth, not to magnetic mountains, as often held at the time;[38] he believed to have demonstrated this effect with a *terrella* that presented a depression on its surface.[39] The present explanation is more complex and is related to the mechanisms acting inside the Earth's core that give rise to the magnetic properties of the planet (Chap. 7).

Some of Gilbert's ideas found opposition in contemporary authors; for example, the Jesuit scholar Athanasius Kircher (1601–1680), an opponent of the Copernican heliocentric theory (enunciated in 1543), did not accept Gilbert's identification of the Earth with the magnet. He thought that if this was the case, the magnetic attraction would be so strong men would not be able to use iron tools; in the book *Magnes, sives De Arte Magnetica* ("The

[32] W. Gilbert (a), op. cit., p. 116.
[33] W. Gilbert (a), op. cit., p. 117.
[34] W. Gilbert (a), op. cit., p. 110.
[35] W. Gilbert (a), op. cit., p. 112.
[36] W. Gilbert (a), op. cit., p. 117.
[37] W. Gilbert (a), op. cit., p. 117.
[38] W. Gilbert (a), op. cit., p. 77.
[39] W. Gilbert (a), op. cit., p. 79.

Magnet, or About the Magnetic Art") (1641) he qualified Gilbert's opinion as "*absurda, indigna et intolerabilis*".[40]

Gilbert presented a theory of the origin of iron, arising from a single element—"earth"; the metallic properties of iron were produced by moist "exhalations." Although he used these Aristotelian terms, he gave them different meanings [26]. Furthermore, he considered iron, rather than silver or gold, the perfect metal, or the "foremost of metals." This is justified both on the grounds of his theory of metals, since iron is "earth in its own nature true and genuine," and also from the countless uses of the metal. Here, one has an interesting example of a consideration of the social function of the metals entering the discussion of mineralogy or chemistry.[41]

Gilbert‘s cosmology does not make an explicit choice in favor of a heliocentric solar system, since he speaks only of the movement of the Earth's rotation, not of the motion around the Sun. However, this choice is consistent with the scheme of the heavens that will be later presented in *De Mundo*.

Gilbert's ideas influenced Johannes Kepler (1571–1630), who considered them one of the three fundamental elements for the formulation of his theory of the orbits of the planets, together with the heliocentric hypothesis of Copernicus (1473–1543) and the astronomical data of Tycho Brahe (1546–1601). "If I believe anything," he wrote, "you after reading my book will be persuaded that I have placed a celestial rooftop upon the magnetical philosophy of Gilbert, who himself has built the terrestrial foundation".[42]

Gilbert's *De Magnete* represented an important step in the direction of creating a new paradigm for the sciences. His systematic use of the experimental method and his critical attitude toward the classics led him to innovate, stimulating many other authors. *De Magnete*, however, had a somewhat archaic language in some parts. For example, it showed in places an animistic inclination: "Wonderful is the loadstone shown in many experiments to be, and, as it were, animate," writes Gilbert.[43] Gilbert often employs a vitalistic language, as, for example, when he speaks of the increasing response of the iron to the attraction of the lodestone, as it gets nearer, describing it as "excited".[44] *De Magnete* also lacks more elaborate quantitative statements, an element that would gain increasing relevance in the evolution of the new scientific method.

---

[40] A. Kircher, *Magnes, sives De Arte Magnetica*, Rome, 1641, p. 479, quoted by M. R. Baldwin [25].

[41] G. Freudenthal (a), op. cit. p. 28.

[42] Johannes Kepler, *Gesammelte Werke*, ed. by Walther von Dyck and Max Caspar, Munich, 1937-75, xvi, 86, Letter to Johann Georg Brengger of 30 November 1607, quoted by M. R. Baldwin [27].

[43] W. Gilbert (a), op. cit., p. 104.

[44] W. Gilbert (a), op. cit., p. 49.

The use of mathematics in the treatment of the matters of physical reality also separated Galileo from the author of *De Magnete.* Speaking of William Gilbert, Galileo wrote: "I have the highest praise, admiration, and envy for this author, who framed such a stupendous concept regarding an object which innumerable men of splendid intellect had handled without paying any attention to it (.) What I might have wished for in Gilbert would be a little more of the mathematician, and especially a thorough grounding in geometry, a discipline which would have rendered him less rash about accepting as rigorous proofs those reasons which he puts forward as *verae causae* [true causes] for the correct conclusions he himself had observed" [28].

Only 12 years before the publication of *De Magnete*, the Great Spanish Armada had been defeated by the English Navy, a historical episode that changed the political map of Europe, and one to which the English cast-iron cannons gave a significant contribution. At that time, only England dominated the technique of the manufacture of such weapons, which substituted the bronze pieces. It is perhaps no coincidence that William Gilbert's book was published at such a time, when England's eminence was connected with the mastering of the nautical techniques and of iron manufacture [29]. It has been pointed out by the Austrian science historian Edgar Zilsel (1891–1944) that a large proportion of *De Magnete* deals with mining and metallurgy, 13% discuss nautical instruments, and 12% discuss navigation in general.[45]

After Gilbert's death in 1603, presumably of the plague [30], his books, stones, and instruments were donated to the Royal College of Physicians; unfortunately, everything was lost when the College was destroyed in the Great London Fire of 1666. Much was also lost from the partial destruction of Colchester, Gilbert's birthplace, when it was under siege during the Second Civil War in 1648 [31].

For his great contribution to magnetism, William Gilbert was honored with the naming of the unit of magnetomotive force, the magnetic analogue to the electromotive force; one gilbert (symbol Gi) measures the magnetomotive force corresponding to 0.75 A in the International System of Units (SI). In the words of the great English poet and dramatist John Dryden (1631–1700):[46]

Gilbert shall live, till *Load-stones* cease to draw,
Or *British* Fleets the boundless ocean awe.

[45] E. Zilsel (a), op. cit., p. 15.

[46] John Dryden, "Poems," ("To My Honour'd Friend Dr Charleton"), quoted by Patricia Fara [32].

Gilbert and the Scientific Revolution

The change in scientific attitude heralded by *De Magnete* merged with other influences and became part of a real scientific revolution, associated with the names of René Descartes (1596–1650), Francis Bacon (1561–1626), and above all, Galileo Galilei (1564–1642).[47] This transformation of the scientific practice and worldview would reach its climax in the second half of the seventeenth century with the English physicist Isaac Newton (1643–1727).

In the early Middle Ages, theology was the queen of the sciences, in the words of St. Augustine (354–430). Augustine thought that should philosophers teach anything "contrary to our Scriptures, that is to Catholic faith, we may without any doubt believe it to be completely false, and we may by some means be able to show this".[48] The gradual separation of the reign of theology and the domain of the sciences, and their demarcation, is associated especially with two medieval philosophers, the Scot John Duns Scotus (c.1266–1308) and the English philosopher and Franciscan priest William of Ockham (c. 1284–1349) [34]; one may say that the intellectual revolution that gave birth to the scientific method had its roots in the works of medieval thinkers.

On the European continent, *De Magnete* influenced many authors, especially Johannes Kepler and Galileo Galilei. However, the importance of the work of William Gilbert was felt more strongly in England, where he was regarded as the founder of the experimental method and also the introducer of Copernican theory [35]. In 1657, the architect and astronomer Christopher Wren (1632–1723), in his inaugural speech as Professor of Astronomy at Gresham College, in London, referred to him as "the father of the new Philosophy".[49]

The Polish astronomer Nicolaus Copernicus (1473–1543) had proposed in his book *De revolutionibus orbium coelestium* ("On the Revolutions of the Celestial Spheres"), published in 1543, that the Earth and the planets turned around the Sun in circular orbits. The fact that the stars did not change position in the sky as the Earth moved around the Sun was justified by Copernicus on the grounds that the stars were much farther away than hitherto admitted. The idea of a much larger universe, with the Earth removed

[47] A. C. Crombie summarizes the contribution of Galileo to the scientific method in the following words: "The special contribution that Galileo's conception of science as a mathematical description of relations enabled him to make to methodology, was to free it from a tendency of excessive empiricism which was the main defect of the Aristotelian tradition, and to give it a power of generality which was strictly related to experimental facts to a degree which previous Neoplatonists had seldom achieved" [33].

[48] Augustine, *De Genesi ad Litteram*, chapter 21, quoted by A. C. Crombie (a), op. cit., vol. 1, p. 75.

[49] J. A. Bennett (a), op. cit., p. 171.

from its privileged position at the center, had a great intellectual impact, starting the scientific revolution. After Copernicus published his heliocentric theory, the decision on its truth became possibly the chief scientific problem of the late sixteenth and early seventeenth centuries [36]. This question occupied the minds of the greatest thinkers of the period, such as Johannes Kepler, René Descartes, and Galileo Galilei.

The scientific revolution gained momentum as men of learning started to study, with a fresh approach, the phenomenon of the movement of the bodies. The first step in this direction was to abandon the attitude of the Greek philosophers, who dealt with motion as a quality essential to the moving bodies, not as a state in which bodies happened to be. The question asked was no longer how to define or characterize the essence of motion, but instead, how to describe motion in mathematical terms. This shift in emphasis occurred in parallel with advances in mathematics, more specifically in geometry, algebra, and mathematical notation; by the first decades of the seventeenth century, algebra and arithmetic had adopted essentially the same notation as is employed today [37].

Galileo Galilei was born in Pisa in 1564 (Fig. 3.5), where he studied medicine; he later became a lecturer at the University of Pisa and in 1592, in Padua. He treated the problem of the motion of bodies, distinguishing between the conditions of uniform velocity and accelerated motion. Experimenting with balls rolling on planes and observing that on smooth

Fig. 3.5 Galileo Galilei (1564–1642), Italian physicist and astronomer who was a pioneer in the mathematical formulation of the laws of physics

planes the motion continued with almost constant velocity, Galileo made the abstraction that in the absence of friction, the velocity would remain constant forever. He concluded, "This particle will move along this same plane with a motion which is uniform and perpetual, provided the plane has no limits" [38]. This was an early formulation of a principle that became known as the principle of inertia; it represented an idea directly opposite to the Aristotelian conception that a body in motion required the constant application of forces.

Galileo was fascinated with the magnetism of the lodestone and conducted many experiments to investigate its properties. He believed that forces similar to magnetic forces acted in space, between the heavenly bodies, and that the Earth was a huge magnet [39]. Galileo read *De Magnete* and was much impressed with its content and methodology; he mentions it in his most complete work, the *Discorsi e dimostrazioni mathematiche intorno a due nuove scienze attenenti alla meccanica* ("Dialogue Concerning Two New Sciences"), published in 1634. In the words of Salviati, who represents Galileo in the dialogues, he would have missed Gilbert's book, "If a famous Peripatetic [Aristotelian] philosopher had not made a present of it, I think in order to protect his library from its contagion".[50] The other character of the Dialogues, Simplicius, asks Salviati: "Then you are one of those people who adhere to the magnetic philosophy of William Gilbert?" and Salviati replies, "Certainly I am, and I believe that I have for company every man who has attentively read his book and carried out his experiments".[51]

However, Galileo also wrote [40]: "I want to tell you about one particular to which I wish Gilbert had not lent his ear," referring with disapproval to the suggestion that Gilbert had made (taken from Peregrinus) that the *terrella* would turn by itself if aligned with the axis of the Earth.

Having heard of the invention of the telescope in Holland, Galileo built one himself in 1609, calling it *perspicillum*; the first version magnified only three times, but as Galileo refined his technique, he produced much better instruments. One may say that he thus built some of the first scientific instruments; his observations with the telescope would revolutionize humankind's understanding of the universe. One can imagine the emotion of Galileo as he pointed his telescope at the night sky for the first time. He was stunned to perceive on the surface of the Moon, previously regarded as smooth, mountains that resembled those on Earth; near Jupiter, satellites that turned around the planet; spots on the face of the sun; thousands of stars never previously imagined, and the image of Venus showing phases like the Moon! About the moon,

[50] Galileo Galilei (a), Third Day, p. 400.

[51] Galileo Galilei (a), Third Day, p. 400.

he wrote:[52] "It is a most beautiful and delightful sight to behold the body of the moon..."

The presence of mountains on the Moon and spots on the Sun's surface helped to destroy the idea of supralunary and perfect spheres, where the stars and the Sun were located, and which differed from the Earth, the realm of corruption and imperfection. This inaugurated a picture of a unified world and meant the end of the idea of Cosmos, or a finite, closed hierarchy of spheres. The importance of this transformation is characterized by the science historian Alexandre Koyré (1892–1964) in strong words: "The dissolution of the Cosmos—I repeat what I have already said: this seems to me to be the most profound revolution achieved or suffered by the human mind since the invention of the Cosmos by the Greeks" [41]. This revolution would have its martyrs, like the philosopher Giordano Bruno (b. 1548), who was condemned by the Inquisition for his philosophical ideas, his heliocentric views, and his belief in an infinite universe and burned at the stake in Rome in 1600.

Galileo too was persecuted by the Church for his teachings on the Copernican worldview and was condemned by the Inquisition and sentenced to spend his last years of life under house arrest. During this period, he was still intellectually very active and published his last book, "Dialogue Concerning Two New Sciences." He was forced to recant his beliefs under the threat of torture: he concluded his statements with the words, "I, Galileo Galilei, have abjured as above with my own hand".[53] Galileo was "rehabilitated" by the Church only 360 years later, in 1992.

René Descartes, another of the founders of the New Science, was born in La Haye (now Descartes), in France, in 1596. He studied law in Poitiers and later lived for many years in the Netherlands. Descartes was one of the chief advocates of the new picture of the universe, where the same laws ruled terrestrial and celestial phenomena. In his book *Le Monde* ("The World"), written in the early 1630s, he attempts a mechanical explanation of the universe, avoiding scholastic concepts. Although Descartes made very important contributions to mathematics and to mathematical techniques used in physics, he expressed his ideas in physics mostly in non-mathematical terms and insisted on the search for intrinsic or essential causes; he remained a philosopher dealing with physical problems. In that approach, he differed from the way Galileo treated the issues of the physical world. For example, commenting on the work of Galileo, he wrote: "As to what Galileo has written about the balance

[52] Galileo Galilei, quoted by Bronowski, p. 204.

[53] Galileo Galilei, quoted by Bronowski, p. 217.

and the lever, he explains very well what happens (*quod ita fit*), but not why it happens (*cur ita fit*), as I have done in my *Principles*".[54]

The mathematization of the worldview, implicit in the method of Galileo, has its roots in the teachings of the Pythagoreans. Plato, who lived from 427 BC to 347 BC, continued this tendency; for him, the structure of the universe was related to mathematical entities in an intimate way: particles of fire were tetrahedra; air was formed of octahedra, and so on. The universe itself had a spherical shape, since the sphere is the most perfect of forms. The point of view of Plato and his followers contrasts with the vision of Aristotle in many fundamental ways, and also in this specific point; Aristotle framed his arguments in qualitative, not quantitative terms.

Adherence to this view usually leads to the choice of mathematics as the language most adequate to express the facts of nature. Galileo Galilei, at the beginning of the seventeenth century, conveys [42] this view in his book *Sidereus Nuncius* (1610): "Philosophy [nature] is written in that great book which ever lies before our eyes. I mean the universe, but we cannot understand it if we do not first learn the language and grasp the symbols in which it is written. The book is written in the mathematical language, and the symbols are triangles, circles and other geometrical figures without whose help it is humanly impossible to comprehend a single word of it, and without which one wanders in vain through a dark labyrinth." This concept reached maturity only with the development of other mathematical techniques used to describe more complex physical processes, such as the differential calculus, invented by Newton and Leibniz in the seventeenth century.

The astronomer Johannes Kepler, born in 1571 in Weil-der-Stadt, Swabia, Southern Germany, solved the problem of the orbits of the planets. Holding the Copernican view that the Earth and the other planets revolved around the Sun, and using the astronomical observations of the Danish astronomer Tycho Brahe (1546–1601), Kepler discovered that the planets described elliptical, not circular orbits. Inspired by the "natural philosophy" of Gilbert, he considered that the planets were maintained in their motion by magnetic forces.

The scientific revolution would complete its cycle with the mechanics and the theory of gravitation of Isaac Newton. The laws of Kepler were then shown to fit into a scheme of forces of gravitational attraction whose intensity fell with the square of the distance. These forces determined the orbits of the planets, satellites, and comets.

[54] R. Descartes, 1638, quoted by A. C. Crombie (a), op. cit., vol. 2, p. 171.

Summing up the accomplishments of science in the seventeenth century, the American astronomer Herbert Dingle (1890–1978) [43] wrote: "The achievement of the seventeenth century had been precise, profound and immeasurably great. At the beginning of that great epoch the conception of the universe which, though not unchallenged, was still predominant, was essentially medieval in spirit; at the centre were the earth and the baser elements, subject to change and decay and tending to move in straight lines towards their own places, while surrounding them the crystalline spheres, incorruptible in the heavens, performed eternally the perfect circular motions proper to their nature. By the end of the century, the whole scheme had vanished, and the Newtonian system reigned in its stead. The spheres had disappeared; matter and motion were the same everywhere, on the earth and in the heavens, and were indissolubly linked with one another through the universal force of gravitation; and a technique had been created for the conquest of other phenomena by means of Newtonian forces yet to be discovered."

References

1. Harré, R. (1970). *The method of science* (p. xi). Wykeham Publications.
2. Beaujouan, G. (1966). La Science dans l'Occident Médiéval Chrétien. In R. Taton (Ed.), *Histoire Générale des Sciences* (Vol. I, p. 583). Presses Universitaires de France.
3. Crombie, A. C. (1995). *The history of science from Augustine to Galileo* (Vol. 2, p. 117). Dover Publications.
4. Crombie, A. C. (n.d.). *The history of science from Augustine to Galileo* (p. 126). Dover Publications.
5. Arnaldez, R., Massignon, L., & Youschkevitch, A. P. (n.d.). La Science Arabe. In R. Taton (Ed.), p. 462.
6. Ronan, C. A. (1984). *The Cambridge illustrated history of the world's science* (p. 229). Cambridge University Press.
7. Schmidl, P. G. (1997–98). Two early Arabic sources on the magnetic compass. *Journal of Arabic and Islamic Studies, 1.*
8. Merrill, R. T., McElhinny, M. W., & McFadden, P. L. (1998). *The magnetic field of the earth* (p. 170). Academic Press.
9. Smith, J. A. (1992). Precursors to Peregrinus: The early history of magnetism and the mariner's compass in Europe. *Journal of Medieval History, 18*, 21.
10. Smith, J. A. (1992). Precursors to Peregrinus: The early history of magnetism and the mariner's compass in Europe. *Journal of Medieval History, 18*, 21, p. 34.
11. Smith, J. A. (1992). Precursors to Peregrinus: The early history of magnetism and the mariner's compass in Europe. *Journal of Medieval History, 18*, 21, p. 37.

12. Majumdar, C. K. (1981). Magnetism: Past, present and future. In N. S. S. Murthy & L. M. Rao (Eds.), *Current trends in magnetism* (p. 1). Indian Physics Association.
13. Smith, J. A. (1992). Precursors to Peregrinus: The early history of magnetism and the mariner's compass in Europe. *Journal of Medieval History, 18*, 21, p. 38.
14. Smith, J. A. (1992). Precursors to Peregrinus: The early history of magnetism and the mariner's compass in Europe. *Journal of Medieval History, 18*, 21, p. 24.
15. Smith, J. A. (1992). Precursors to Peregrinus: The early history of magnetism and the mariner's compass in Europe. *Journal of Medieval History, 18*, 21, p. 56.
16. Thorndike, L. (1945). John of St. Amand on the magnet. *Isis, 36*, 156–157.
17. Grant, E. (1980). Peter Peregrinus. In C. C. Gillispie (Ed.), *Dictionary of scientific biographies*. Charles Scribner's Sons.
18. de Maricourt, P. (1974). The letter of Peregrinus. In E. Grant (Ed.), *Source book in medieval science* (p. 368). Harvard University Press.
19. King, W. J. (1959). The natural philosophy of William Gilbert and his predecessors. *Contributions from the museum of history and technology. Smithsonian Institution Bulletin 218*, 122–139, p. 127.
20. Bernal, J. D. (1972). *The extension of man* (p. 124). Paladin.
21. Zilsel, E. (1941). The origins of William Gilbert's scientific method. *Journal of the History of Ideas, 2*(1), 1, p. 23.
22. King, W. J. (1959). *The natural philosophy of William Gilbert and his predecessors*, Contributions from the Museum of History and Technology. *Smithsonian Institution Bulletin 218*, 122–139, p. 124.
23. Gilbert, W. (1978). *De Magnete, Great books* (Vol. 28, p. 1). Encyclopaedia Britannica.
24. Roller, D. H. D. (1959). *The De Magnete of William Gilbert* (p. 92). Menno Hertzberger.
25. Baldwin, M. R. (1985). Magnetism and the anti-Copernican Polemic. *Journal of the History of Astronomy, 16*, 155, p. 159.
26. Freudenthal, G. (1983). Theory of matter and cosmology in William Gilbert's De magnete. *Isis, 74*, 22–37, p. 25.
27. Baldwin, M. R. (1985). Magnetism and the anti-Copernican Polemic. *Journal of the History of Astronomy, 16*, 155, p. 156.
28. Galilei, G. (1967). *Dialogues concerning the two chief world systems—Ptolemaic & Copernican*, "Third Day" (S. Drake, Trans., 2nd ed, p. 406). University of California Press.
29. Zilsel, E. (1941). The origins of William Gilbert's scientific method. *Journal of the History of Ideas, 2*(1), 1, p. 31.
30. Roller, D. H. D. (1959). *The De Magnete of William Gilbert* (p. 91). Menno Hertzberger.
31. Roller, D. H. D. (1959). *The De Magnete of William Gilbert* (p. 50). Menno Hertzberger.
32. Fara, P. (1996). *Sympathetic attractions*. Princeton University Press.

33. Crombie, A. C. (1953). *Robert Grosseteste and the origins of experimental science 1100–1700* (p. 305). Clarendon Press.
34. Ronan, C. A. (1984). *The Cambridge illustrated history of the world's science* (p. 260). Cambridge University Press.
35. Bennett, J. A. Cosmology and the magnetical philosophy, 1640–1680. *Journal of the History of Astronomy, 12*, 165, p. 166.
36. Crombie, A. C. (1995). *The history of science from Augustine to Galileo* (Vol. 2, p. 145). Dover Publications.
37. Crombie, A. C. (1995). *The history of science from Augustine to Galileo* (Vol. 2, p. 138). Dover Publications.
38. Galilei, G. (1978). *Dialogues concerning the two new sciences.* Fourth Day, Great Books (Vol. 28, p. 238). Encyclopedia Britannica.
39. Shea, W. R. (1977). *Galileo's intellectual revolution* (2nd ed., p. 167). Science History Publications.
40. Galilei, G. (1967). *Dialogues concerning the two chief world systems—Ptolemaic & Copernican,* "Third Day" (S. Drake, Trans., 2nd ed., p. 413). University of California Press.
41. Koyré, A. (1943). Galileo and Plato. *Journal of the History of Ideas, 4*(4), 400, p. 404.
42. Kline, M. (1985). *Mathematics and the search of knowledge* (p. 95). Oxford University Press.
43. Dingle, H. (1972). Physics in the eighteenth century. In A. Ferguson (Ed.), *Natural philosophy through the eighteenth century* (p. 28). Taylor and Francis.

Further Reading

Aczel, A. D. (2001). *The riddle of the compass.* Harcourt.
Crombie, C. (1995). *The history of science from Augustine to Galileo.* Dover Publications.
de Maricourt, P. (1974). *The letter of Peregrinus, in source book in Medieval Science,* E. Grant (Ed.), Harvard University Press.
Gilbert, W. (1978). *De Magnete, Great books* (Vol. 28). Encyclopaedia Britannica (Original work published 1600).
Pumfrey, S. (2002). *Latitude & the magnetic earth.* Icon Books.
Roller, D. H. D. (1959). *The De Magnete of William Gilbert.* Menno Hertzberger.
Westfall, R. S. (1993). *The life of Isaac Newton.* Cambridge University Press.

4

A Very Feeble Effect: Magnets and Electricity

From time immemorial, as long as there has been any natural science, its ultimate supreme goal has been the combination of the motley diversity of physical phenomena into a unified system or even a single formula; (...) [1]
Max Planck, in "The Unity of the Physical World-Picture."

Summary The Roman naturalist Pliny the Elder (23–79 AD) acknowledged the similarity between the attraction of the magnet and the attraction of pieces of paper by amber rubbed by silk or wool. This parallel suggested some link between the two phenomena. In 1820, Hans Christian Oersted (1777–1851) proved it, observing that an electric current created a field that moved a compass needle. André-Marie Ampère (1775–1836) showed that a coil of copper wire, if free to turn, behaved as a compass needle. Michael Faraday (1791–1867) studied the symmetric effect: he obtained an electric current with a magnet moving in and out of a coil, inventing the first generator. Many investigators built electrostatic machines that used friction to charge different objects. Until the eighteenth century, it was not known that the electrification by friction and current transport through conductors involved the same charge carriers. The hypothesis of one single electrical fluid was made by Benjamin Franklin (1706–1790). André-Marie Ampère (1775–1836) suggested that the properties of the magnet were derived from "molecular" currents inside it. Alessandro Volta (1745–1827) created the first electric battery by piling several pairs of discs of two different metals, separated by cloth or cardboard moistened in brine.

A. P. Guimarães, *A Longstanding Attraction*, https://doi.org/10.1007/978-3-032-02006-2_4

Electric Shock

Since antiquity, two instances of action at a distance were known: the attraction of dust and chaff by a piece of amber rubbed by cloth or fur and the more remarkable magnetic attraction of iron by the lodestone. In his *Natural History*, written in the first century AD, Pliny the Elder (23–79 AD) acknowledges in his notes the similarity between the two effects, since amber [2] "attracts straw, dry leaves and bark from the linden-tree, just as a magnet attracts iron."

The first phenomenon has an electrical origin; amber and the cloth, initially uncharged, acquire with rubbing charges of opposite signs: one develops an excess of electrons (particles of negative charge) and the other an electron deficiency (therefore, a positive charge).

Amber was well known in Greece since the ninth century BC [3], and from *elektron*, the Greek word for amber, derives the word electricity. A charged piece of amber induces an electric charge of the opposite sign on the nearest surface of the initially neutral chaff, and therefore, the chaff is attracted by the amber since opposite charges are attracted and like charges are repelled. The presence of an electrically charged object can be detected using this property of repulsion of equal charges: the two halves of a thin folded leaf of paper or gold separate when in contact with a charged object. An instrument made with such a leaf enclosed in a glass jar is known as an electroscope and has evolved from a pair of linen threads first used for this purpose by Benjamin Franklin (1706–1790) in the eighteenth century.

The process of charging objects by friction relies on the possibility of displacing the elementary negative particles, which were identified in the nineteenth century with the electrons. This displacement is also at the basis of the charge transport in electrical conductors, like the copper wires of the electricity network. Until the eighteenth century, it was not evident that the effects of electrification by friction, the so-called static electricity effects, and the transport of current through conductors both involved the same charge carriers.

Another phenomenon of electric origin known in Greek and Roman antiquity is the power of some fish to induce numbness (*torpere*, in Latin); among these are the rays called torpedoes (of the family Torpedinidae), of which several species occur in the Mediterranean (Fig. 4.1). Plato refers to the electric ray in the dialogue where Meno tells Socrates that his "power over others to be very like the flat torpedo fish, who torpifies those who come near him and touch him" [4]. Aristotle, in his *History of Animals*, also described that the

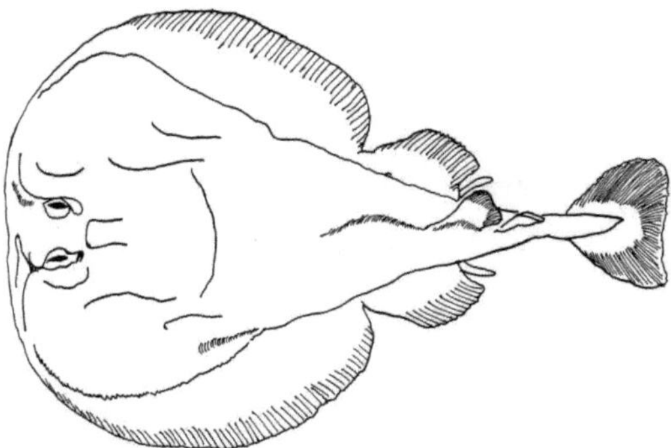

Fig. 4.1 The electric ray *Torpedo nobiliana*, used in Greece and Rome for the treatment of several diseases

torpedo "narcotizes the creatures that it wants to catch, overpowering them by the power of shock that is resident in its body, and feeds upon them" [5]. The therapeutic use of these fish to cure headaches was recommended by Scribonius Largus (first century AD), a Roman physician, who published about the year AD 47 the *Compositiones medicamentorum* [6]. Only in the second half of the eighteenth century was the electrical character of the action of the electric rays discovered, after some studies of the South American electric eel Electrophorus electricus [7].

The first systematic study of the electrical properties of materials was published by William Gilbert (1544–1603) in his book *De Magnete* (1600), a treatise mostly devoted to magnetism (Chap. 3). He made an important classification of substances according to their electric behavior; he determined which materials showed the same effect of electric attraction as amber and called them *electrics*. In this class of substances, he included glass, crystal, diamond, sapphire, sealing wax, and sulfur; the other class of materials, the *non-electrics*, for Gilbert, included the metals, for example.

Another earlier researcher of electricity was the Italian Jesuit Niccolo Cabeo (1586–1650), who published in 1629 the book *Philosophia Magnetica*. In this work, the fact that bodies with electric charges of the same polarity show a force of repulsion was enunciated for the first time: "filings attracted by excited amber sometimes recoiled at a distance of several inches after making contact".[1] This phenomenon is due to the fact that the filings, once in contact with the amber, acquire a charge of the same sign, therefore the force of repulsion. This effect was discussed by the German physicist Otto von Guericke (1602–1686)

[1] Niccolo Cabeo, "Philosophia Magnetica" (1629), vol. II, p. 21, quoted by A. Wolf, *History of Science, Technology, and Philosophy in the 17th Century*, p. 303.

in Book IV of his *Experimenta Nova (ut vocantur) Magdeburgica de Vacuo Spatio* ("New Magdeburg Experiments on Void Space"), published [8] in 1672. The English physicist Isaac Newton (1642–1727) also observed, around 1675, electric repulsion when he rubbed a piece of glass, close to some pieces of paper [8].

The British chemist Stephen Gray (1666?-1736) discovered in 1729 that by rubbing an object, he could make another far away object attract light bodies, provided they were connected through some adequate materials; this effect proved the existence of electric conduction. Gray also discovered the induction of charges: a piece of lead hanging from a thread was electrically charged by approaching, without physical contact, a charged glass tube. These experiments were repeated in Paris by Charles-François Du Fay (1638–1739), who published his research in 1733. He is the first person to correlate the two categories of materials created by Gilbert—*electrics* and *non-electrics*—with their degree of electrical conductivity [9]: the *electrics* were the materials that did not conduct electricity—the insulators—and the *non-electrics* were the conductors, or, in his words, materials that "transmitted the electric matter" [9]. The denomination of "conductors" would be used later by Jean Théophile Desaguliers [10] (1638–1744). The non-conductors were called by Desaguliers "electrics per se," or "supporters."

From the analysis of attraction or repulsion between different electrified materials, Du Fay concluded that there exist two different types of electricity: "resinous electricity" and "vitreous electricity." In modern terms, this meant that there are two classes of materials: some that, after being rubbed, have an excess negative charge (resinous) and others an excess positive charge (vitreous). From this distinction, the picture of electric conductivity in terms of the flow of *two* electric fluids resulted naturally, an idea that was for a long period adopted by the investigators and is now known to be incorrect. Du Fay reported that he had observed vitreous electricity in glass, rock crystal, precious stones, and wool, and resinous electricity in amber, silk, and paper. However, other observers later found that depending on which material the vitreous or the resinous materials were rubbed with, they could show either type of electricity.

In the following years, in the same century, it became clear that only one electric species (known today to be the negatively charged electrons) is transported in the phenomenon of electric conduction. This one-fluid hypothesis was first formulated by the American scholar and political leader Benjamin Franklin (1706–1790), who became famous for his experiments with atmospheric electricity made around 1750. Franklin found that the electric effects were due to the excess or deficiency of the electric fluid, also known at the time as "electric matter," or "electric fire." He demonstrated this fact [11] by charging two men standing on glass (insulating) stools; one was charged plus and the other minus. When they touched hands, there was a spark, since the charges flowed from one to the other, re-establishing neutrality in both. If either touched a third uncharged man, they would feel electric shocks, since electricity would flow to the one that had a deficiency of charge, or from the one that had an excess of it.

Franklin also explained that electric charge, or "electric matter," was not *created* by rubbing objects; it was simply displaced from one object to the other. The charging of objects by rubbing is observed when the electric charges leave the rubbed object (glass, sealing wax, or any other electrically insulating material) to the cloth, or vice versa. This and other related phenomena of electrical origin could not be studied in a more systematic fashion until efficient forms of charging objects and ways of storing the electric charge had been devised. Until this happened, observers were practically limited to natural sources of electricity. Among these sources in nature, the phenomena related to the accumulation and flow of atmospheric charge stood out.

Lightning has been known for its frightening effects since the early days of the prehistory of humankind. This phenomenon arises from the accumulation of charges in the clouds, which may lead to huge differences of potential, of the order of a hundred million volts, either between different parts of the clouds or between the clouds and the ground. Lightning is a very intense electrical discharge observed during thunderstorms that tends to equalize these potentials, with currents typically reaching 10,000 A and lasting less than one-tenth of a second. This is a reasonably common phenomenon; at each given instant, some 2000 thunderstorms occur at different points of the Earth.

Although everyone is accustomed nowadays to attributing these natural phenomena to atmospheric electricity, before the beginning of the eighteenth century, the fact that thunder and lightning had electrical origins was not known. In 1708, the English investigator William Wall produced sparks by sliding a long piece of amber through a woollen material and compared the appearance of the sparks and the associated sounds to those of atmospheric phenomena [12, 13]. Benjamin Franklin was the first to definitely establish this connection. In his famous experiment with atmospheric electricity (1752), Franklin managed to use the line of a kite to conduct the electric charges from the atmosphere, which were then safely led to the ground; his studies resulted in the invention of the lightning rod.

The historical records are full of events involving atmospheric electricity, including many tragedies produced by lightning. Some observers attempted to study this force of nature more closely and paid with their lives for their curiosity. One example is that of the German physicist Georg Wilhelm Richmann (1711–1753), killed in St. Petersburg in 1753 by lightning while studying atmospheric electricity. He had set up a conductor on the roof of his house and was trying to make measurements with an electroscope during a thunderstorm [14].

The need for sources of electric charges for the accomplishment of electricity experiments led many investigators to build machines that used friction to charge different objects. Many electrostatic machines were built; one of the first such machines was invented by the German physicist Otto von Guericke (1602–1686) in 1663. Von Guericke is best known for the invention of the air pump and for the experiments with evacuated vessels. He once evacuated two metallic hemispheres (the Magdeburg hemispheres) and showed that atmospheric pressure held them with such a force that they could not be separated by a team of 16 horses.

Von Guericke's electrostatic machine consisted of a globe of sulfur on an iron axis around which it could turn. Rubbing the turning globe with his hands or with a piece of cloth, von Guericke charged the globe and managed to make several experiments with electricity. He observed that the charged globe could induce a charge on other objects and also that electricity could be carried by some bodies. The conduction of electricity was demonstrated by showing that electric effects, such as the attraction of small objects, would seem as if far-away objects were physically connected to the charged globe.

In 1675, Jean Picard (1620–1682), a French astronomer at the Collège de France, the first to measure accurately the length of a degree of a meridian, made an interesting observation. He noticed, as he transported during the night, a barometer consisting of a glass tube filled with mercury, the appearance of a blue light inside the tube [15]. This phenomenon was investigated systematically 30 years later by the English self-taught scientist Francis Hauksbee, the Elder (c. 1666-c. 1713); Hauksbee found then that the luminosity was due to static electricity produced by the friction of the mercury against the glass tube. Inspired by this discovery, he made an electrostatic generator consisting of a turning glass tube that was rubbed by his hands; he had reinvented von Guericke's electrostatic machine, using glass instead of sulfur.

Another significant advance in the study of electrical phenomena was the accidental discovery of a suitable means of storing electrical charge—the Leyden jar—in Holland in 1746 by the Dutch physicist Pieter van Musschenbroek (1692–1761). This device was simultaneously invented by the administrator and cleric Ewald Georg von Kleist (c. 1700–1748) from

Pomerania in Prussia. The Leyden jar consisted initially of a glass jar filled with water, with a conducting cable immersed in it. This original design was later improved into a glass bottle lined with metal foil, both internally and externally. It works by storing opposite charges in the two metallic foils. The Leyden jar is a direct ancestor of the present-day capacitor, a common component in any electronic circuit.

Many experiments in electricity were made by combining electrostatic machines with Leyden jars. In 1847, William Watson (1715–1787) and other fellows of the Royal Society managed to send an electric signal across the Thames, discharging a Leyden jar through the water of the river, by inserting two wires, one on each bank, 400 yards apart [16].

Leyden jars were used to produce electrical shocks in human subjects; this unpleasant sensation arises from the flow of electricity through the body tissues. Depending on the intensity of the current and its path, electrical shocks may, of course, be lethal. Musschenbroek himself, reporting his discovery to the French physicist René Antoine de Réaumur (1683–1757) in 1746, wrote [17]: "I wish to report to you a new but terrible experiment, which I advise you on no account to attempt yourself."

In France, the electrical capacity of the Leyden jars was demonstrated in many ways; on one occasion, King Louis XV witnessed the discharge of a jar through a human chain of 180 Royal Guards holding hands. Many amateurs exhibited the remarkable effects of electricity in the eighteenth century European salons.

The Electric Battery

A more effective form of inducing electrical phenomena had to wait until electricity was produced by chemical reactions. The first indication that electricity was made to flow as a consequence of these reactions was the series of experiments conducted by the Italian anatomist Luigi Galvani (1737–1798) (Fig. 4.2) in the 1780s in Bologna.

The physiological effects of electric shocks, including the acceleration of the pulse, perspiration, and contraction of the muscles, led to the picture of electricity as a "vital force." The resuscitation, or "reanimation," of drowned or apparently dead people was proposed in 1778 by Charles Kite at the Royal Human Society in London; those who did not respond to these stimuli were considered dead [18].

Galvani started in 1780 a study of the response of the nerves and muscles of frogs to static electricity. He connected an electrical machine to the

Fig. 4.2 The Italian anatomist Luigi Galvani (1737–1798)

assembly of the spinal cord, crural nerve, and lower limbs of the frogs and observed that the legs contracted when subjected to electric discharges. Then he found out that the legs contracted even when insulated from the machine whenever the machine produced a spark. He tried the same experiment using atmospheric electricity and again obtained a contraction.

An unexpected effect was observed when he hung the legs with brass hooks from an iron railing in the garden of his house. Galvani had observed that the legs contracted during a thunderstorm, but this time he found that they still twisted when the storm was over, when "the sky was quiet and serene" [19]. They showed the same behavior when taken indoors; the response was observed every time the leg was in contact with two dissimilar metals. As he touched the limbs with different metallic objects, Galvani noticed that the intensity of the response depended on which metals were employed.

Galvani (incorrectly) interpreted these observations as proof of the existence of some "animal electricity," "un'elettricità particolare": he thought that the muscles of the dead animal stored an electric fluid produced by the brain that was responsible for these effects. When his observations were published in 1791 in the paper *De viribus electricitatis in motu musculari commentarius* ("Commentary on the Effect of Electricity on Muscular Motion"), they produced a strong impact. Galvani was then a professor of obstetrics at the Istituto delle Scienze at the University of Bologna.

Fig. 4.3 The Italian physicist Alessandro Volta (1745–1827), inventor of the electric battery

Among those who reacted to these discoveries was the Italian researcher Alessandro Volta (1745–1827) (Fig. 4.3); he did not have a high opinion of the physicians, whom he considered "ignorant of the known laws of electricity" [20]. Volta correctly explained the results of Galvani, pointing out that they were due to the chemical reaction between the two metals through the moisture covering the frog legs. In present-day words, one might say that when two metals are separated by a conducting liquid (called an electrolyte), a potential difference appears between them, an electric current flows, and this stimulates the nerves and then the muscles in the frog limbs. The electrical potential had nothing to do with "animal electricity"; only its detection had to do with the sensitivity of the nerves of the frog to electrical stimulation.

The identification of electricity with the 'vital force,' or its connection to the essence of life, remained in the imagination of authors throughout the eighteenth and nineteenth centuries. An expression of the forcefulness of this link is given by the novel *Frankenstein, or the Modern Prometheus* (1818), by the English novelist Mary Godwin Shelley (1797–1851). The suggestion made by the author is that "Perhaps a corpse would be reanimated; galvanism had given token of such things; perhaps the component parts of a creature might be manufactured, brought together, and endued with vital warmth".[2] *Frankenstein* describes how the experiment of creating a humanoid from parts

[2] M. Godwin Shelley, *Frankenstein or the Modern Prometheus*, edited by M. K. Jozepth, Oxford University Press, Oxford, 1969, quoted by Laura Bossi [21].

of corpses takes a terrible turn after life is instilled into the monster by the application of a powerful electric discharge.

In our age, the link binding electricity to life is present in the technique of electric defibrillation: electric shocks are used to save the lives of patients whose hearts engage in an erratic electrical activity called fibrillation. The electric activity of living tissue is also at the basis of several useful diagnostic procedures; for example, electromyography (EMG) studies muscle activity, electroencephalography (EEG) studies activity of the brain, and electroretinograms (ERM) study the electric potentials associated with the stimulation of the retina (see Chap. 8).

Alessandro Giuseppe Antonio Anastasio Volta was a physicist, born in Como, Italy, in 1745. When he was 30, he invented the electrophorus, a static electricity generator formed of a disc made of resin and wax that, when rubbed, develops an electric charge; a metallic disc closely mounted acquires an induced charge. Volta conducted many experiments in electricity, and his most important contribution came when he was a professor of experimental physics in Pavia in the 1790s. He was a sociable man and had other interests, judging from the comment of his friend Lichtenberg, according to whom Volta "understood a lot about the electricity of women" [22].

At the beginning of his search after the announcement of the discovery of Galvani, Volta took up the experiments of the anatomist "with little hope of success" [20], but in April 1792, he was able to repeat the same results, giving them the correct interpretation.

In the experiments with the frogs, the potential difference was detected by the nerves; Volta invented another method of detection. He found that two discs of dissimilar metals in contact with his tongue gave a strange feeling; saliva was the intervening medium in this experiment. He did not know that this effect had been reported in 1762 by the Swiss Professor of Mathematics Johann Georg Sulzer (1720–1779), who, however, had not related it to electricity [23].

We now know that when two pieces of metal are separated by an electrolyte, there is a chemical reaction that gives rise to a small potential difference, e.g., in the case of zinc and silver, of 0.78 V. To multiply this potential difference, Volta piled several pairs of discs of these two metals, each pair separated by a piece of cloth or cardboard moistened in brine. This arrangement became known as the voltaic pile and constituted the first electric battery. When the two outermost discs were connected through a conducting wire, a flow of electric current was observed. This discovery was communicated in a letter in 1800 to Sir Joseph Banks, president of the Royal Society of London; it was published in the *Philosophical Transactions* of the Society in the same year.

Alessandro Volta demonstrated his discoveries at the Academy of Sciences of Paris, and they made a strong impression on Napoleon, who decided to award a gold medal "for the best experiment made each year on the galvanic fluid [electricity]", and a prize of 60,000 francs for contributions comparable to "Franklin's and Volta's" [24].

Volta's invention created the means of generating steady electric currents, which opened up a new world of experimental possibilities with electricity, explored in subsequent years, especially with the work of Oersted, Ampère, and Faraday.

Volta's name was perpetuated in the name of the unit of potential difference, or tension, in the International System, the volt (abbreviated V).

It was finally established during the eighteenth century that frictional electricity, or static electricity, was the same as "galvanic" electricity, or the electricity produced by Volta's pile. To understand why these phenomena sometimes present a different form, one has to examine more carefully the meaning of the terms "voltage" or "potential difference" and "current" To understand these concepts, relevant to the study of electricity, a parallel is usually made between the electric circuit and a system of water pipes. The voltage or potential difference in electricity is analogous to the potential energy of a water reservoir placed at a certain height; the intensity of the electric current (or amount of charge circulating per unit time) is analogous to the water output (volume of water per unit time) flowing down from the reservoir through the pipe. In frictional or static electricity, the experiments usually involve high potential differences and small currents; sparks are only observed when the potential differences are high, of the order of hundreds or thousands of volts. The currents last only a short time, and the total charges transported in these experiments are usually small. The opposite is true of the effects related to the voltaic pile, which provided lower voltages and steady currents.

In an electric circuit, the intensity of the current is related to the voltage. For a given circuit, characterized by a certain resistance, the current is directly proportional to the voltage across it and inversely proportional to the value of this resistance. This relationship, valid for most materials, is known as Ohm's Law, after the German physicist Georg Simon Ohm (1787–1854), a schoolteacher in Cologne who published it in 1827.

The Unity of Nature

Nature presents to the eye of the inquisitive observer a wide variety of processes and objects of study: the light-emitting Sun, the diversity of landscapes on the Earth, the Moon and the stars that decorate the night sky, the wandering planets, and so on. Several centuries before our era, Presocratic philosophers in Greece looked for a unity in nature underlying the apparent wide diversity of natural phenomena. The existence of this unity, according to the early Greek thinkers, was intertwined with the concept of the universe as an organism [25]. This idea of unity is particularly clear in the teachings of the Pythagoreans.[3] Instead of unity, one can also speak, in line with these thinkers, of a kinship of nature [27].

Unity in this context is the unity in nature, not in science; in other words, the unity of the object of study of the sciences, something different from the unity in the corpus of science, posited by some philosophers of science. One may argue, however, that the unity of nature would imply unity in the sciences. In the eighteenth century, Roger Boscovich (1711–1787) (or Rudjer Josip Boscovic), a Croatian Jesuit born in Ragusa, present-day Dubrovnik, published in Vienna (in 1758) his *Theoria Philosophiae Naturalis* ("Theory of Natural Philosophy"), and his project was to "derive all observed physical phenomena from a single law".[4] This is the same aim modern physicists pursue under the over-ambitious name of "Theories of Everything." One of these theories is the string theory of the particle physicists, which attempts to describe in a unified way all the physical interactions (or "forces"), including gravitation and electromagnetism [29]. These strings are elementary entities that can vibrate, and one can, in a way, think that the "notes" of these vibrations are the particles, such as the electrons. However, there are many problems with this theory, one of them being that it requires a universe of ten dimensions, with seven of these practically inaccessible to experiment.

In an attenuated form, the goal of attaining a unified picture of the physical world is shared by all physicists. The German physicist Carl Friedrich von Weizsäcker (1912–2007), known as a proponent of a theory of the formation of the solar planets, considers "the endeavor of physics to achieve a unified world-view. We do not accept appearances in their many-colored fullness, but

[3] According to Guthrie: "The world is divine, it is therefore good, and is a single whole. If it is good, alive and a whole, that is because, said Pythagoras, it is *limited*, and displays an *order* in the relations of its various parts. Full and efficient life depends on organization" W. K. C. Guthrie [26].

[4] Roger Boscovich, *Theoria Philosophiae Naturalis*, quoted in J. D. Barrow [28].

we want to explain them, that is, we want to reduce one fact to another".[5] A similar view, given in the opening quotation of this chapter, was expressed in the twentieth century by the German physicist Max Planck, the founder of quantum mechanics, the physics that describes the atomic and subatomic world.

German *Naturphilosophie* (Natural Philosophy), a tendency among some philosophers, rather than a school of thought, flourished in the early nineteenth century and is associated with the name of the philosopher Friedrich Wilhelm Joseph von Schelling, born in 1775 in Leonberg, Germany. He wrote *Ideas on a Philosophy of Nature* (1797), where he exposed his conception of the subject.

Schelling was influenced by the ideas of the philosopher Immanuel Kant (1724–1804) and posited an essential unity in nature. Kant, probably the greatest philosopher of modern times, had written that "(...) all natural philosophy consists in the reduction of given forces apparently diverse to a smaller number of forces and powers sufficient for the explication of the former".[6] This view was relevant in the beginning of the nineteenth century and was related to the German Romantic movement. It had a great influence in biology; in physics, it stimulated the search for relationships between different physical phenomena, including gravitation, electricity, and magnetism.

Before the work of William Gilbert, in the sixteenth century, the identification of electric and magnetic attractions was common; *De Magnete* clearly separated these effects. However, the fact that amber attracted pieces of paper, in an apparently analogous way to the attraction of iron by a magnet, suggested to many people that there existed something in common, or a link, between the two classes of phenomena.

Some isolated facts documented before the nineteenth century indeed pointed to this link between electricity and magnetism. One of them was the "strange effect of Thunder upon a Magnetick Sea-card," reported in 1676.[7] A ship hit by lightning at the latitude of the Bermudas had her foremast broken; she then started to sail in the opposite direction! When the captain of an accompanying vessel changed course and reached the damaged ship, he verified, in the words of the report, that "the card was turned round, the North and South points having changed positions."

---

[5] C. F. von Weizsäcker, "*The World-View of Physics*," 1952, translated by Marjorie Grene, from the 4th German edition, 1949, pg. 30, quoted by W. K. C. Guthrie [30].

[6] Immanuel Kant, *Metaphysical Foundations of Natural Science*, Bobbs-Merrill, Indianapolis, 1970, p. 93, quoted by Timothy Shanahan [31].

[7] *Philosophical Transactions*, London, vol. XI, pp. 647-653, (1676) p. 647.

Another report informed that some cutlery became magnetized when its container was struck by lightning in Wakefield, England, in the year 1731; the discharge had even melted some knives and forks. When the knives were put on a table near some nails, they were attracted. A certain Dr. Cookson, who witnessed this strange phenomenon, speculated that it might have resulted from cooling the molten cutlery in the Earth's field.[8]

The interest and the continued investigation of the unity of electricity and magnetism ushered in 1820, a discovery that was to represent a significant milestone in science. This was the result of the investigation of a Danish physicist, Hans Christian Oersted (1777–1851), on the connections between these two classes of phenomena.

The Little Hans Christian and the Great Hans Christian

One of the most important sights in the city of Copenhagen is the bronze statue of the Little Mermaid, at the entrance of the port. The Little Mermaid is a creation of the Danish writer Hans Christian Andersen (1805–1875), who lived in the first half of the nineteenth century and wrote the famous *Tales of Andersen*. He is regarded as the founder of modern children's literature, creating an innovative language and producing masterpieces that appealed to both children and adults. Andersen is the author of *The Ugly Duckling*, *The Emperor's New Clothes* and *The Red Shoes*, among other books.

This genre had been pioneered in the seventeenth century with the collection known as the *Tales of Mother Goose*, by the French Charles Perrault (1628–1703); these included classics such as *Cinderella* and *The Sleeping Beauty*. In the nineteenth century, Andersen had been preceded by the German brothers Jacob and Wilhelm Grimm (1785–1863 and 1786–1859), authors of *Hansel and Gretel* and *Snow White and the Seven Dwarfs*.

A contemporary, compatriot, and namesake of Andersen, Hans Christian Oersted (1777–1851) (Fig. 4.4), left us no fairy tales, but instead demonstrated beyond doubt the intimate connection between electricity and magnetism.

When the future writer Hans Christian Andersen arrived in Copenhagen, he was a poor boy of 14; he met Oersted and was helped by him, already an adult. They developed a close relationship, and in his recollections, Andersen wrote: "(…) his home became very early a home for me; his children, when

[8] *Philosophical Transactions*, London, vol. XXXIX, (1735) pp. 74 and 75.

Fig. 4.4 The Danish scientist Hans Christian Oersted (1777–1851), discoverer of the effect of an electric current on a compass needle

they were small, I have played with, seen them grow up and keep their love for me. In his home, I have found my eldest and unchanged friends".[9] Once a week, Andersen dined at the Oersteds and continued this routine for 24 years after Hans Christian Oersted's death in 1851. Hans Christian Andersen referred to himself as "Little Hans Christian" and to Oersted as "Great Hans Christian." And great he was, being nowadays remembered as the author of one of the most important experiments in the history of science.

Oersted was born in Rudkoebing, Denmark, in 1777, the son of the owner of an apothecary shop. At the age of 11, he began to work as his father's assistant in the pharmacy; from this experience, he derived a practical knowledge in chemistry and an interest in the sciences in general. In 1794, the family moved to Copenhagen, and after 3 years, Oersted obtained a degree in pharmacy at the university.

Hans Christian Oersted had a strong interest in philosophy and became a follower of Kantian ideas. In 1799, he obtained the degree of Doctor of Philosophy, with a thesis on Kant, *Dissertatio de forma Methaphysices elementaris naturae externa* ("Dissertation on the Elementary Metaphysical Forms of External Natures"). This reflected the influence of Kant's ideas expressed in the *Critique of Pure Reason* and in the *Metaphysical Foundations of Natural Science* [33]. Oersted was a member of the editorial staff of the journal

[9] Hans Christian Andersen, quoted by B. Dibner [32].

Philosophisk Repertorium for Faedrelandets Nyeste Litteratur, devoted to Kantian philosophy.

In the summer of 1801, Oersted started a tour through Europe, visiting several academic and scientific institutions. He visited laboratories in Berlin, Göttingen, and Weimar. He attended lectures on *Naturphilosophie* in Berlin and met some philosophers of that movement, including Johann Fichte and August Schlegel. Although influenced by these ideas, Oersted remained closer to Kant's system and was critical in relation to Natural Philosophy for its speculative attitude towards research; he valued the writings of its followers for their "great beauty" but criticized them for underestimating the importance of observation and experimentation [34]. However, the relevance of Naturphilosophie as an inspiration for Oersted should not be underestimated [35].

Like many other scientists, Oersted interpreted Kant's ideas as implying that the patterns or laws found in natural phenomena were in some sense imposed by the human mind; since God had made man in His image, human reason somehow corresponded to the intentions of the Creator. This idea is vividly expressed by the twentieth-century physicist Arthur Eddington (1882–1944) in the words he used to describe humankind's quest for understanding nature:[10] "We have found a strange footprint on the shores of the unknown. We have devised profound theories, one after another, to account for its origin. At last, we have succeeded in reconstructing the creature that made the footprint. And Lo! It is our own."

Finally, Oersted returned to Denmark in 1804 and was appointed professor of physics and chemistry at the University of Copenhagen in 1806 and full professor (*professor ordinarius*) in 1817. He started his investigations on the interrelation of electricity and magnetism in 1813. At that time, he thought that by confining an electric current to a thin wire, he could produce heat, and by narrowing the wire still further, he would obtain magnetism [37].

While in Weimar, Oersted met the German chemist Johann Wilhelm Ritter (1776–1810), the discoverer of ultraviolet rays and of many relevant physical phenomena. Side by side with his scientific interest, Ritter cultivated astrology and tried to correlate the maximum inclination of the ecliptic with the occurrence of discoveries in electricity. In a letter to Oersted, he predicted another outstanding discovery in the last third of the year 1819 or 1820 [38]. Indeed, in 1820, Oersted discovered that when approaching a compass to a wire carrying an electric current, the compass needle moved. The motion of the needle indicated that there had appeared a magnetic field perpendicular to

[10] Arthur Eddington, quoted by M. Kline [36].

the direction of the wire, aligned tangentially to a circle around it. This observation was a turning point in the history of electricity and magnetism: these two distinct and mysterious areas of inquiry were intimately connected!

According to a letter written by one of his associates, Christopher Hansteen (1784–1873), Oersted had accidentally observed this effect in a classroom demonstration. However, this version of the discovery is disputed nowadays because Oersted had been studying the connection between electricity and magnetism for several years. It is interesting to note that, guided by a priori expectations on the symmetry of cause and effect, Oersted had apparently tried for several years to observe the effect of electric currents by arranging the compass needle perpendicular to the conductor. However, only when he set the needle parallel to it did he succeed in producing a noticeable effect.[11]

The observed effect was described by Oersted in the following words:[12] "The magnetical needle, although included in a box, was disturbed; but as the effect was very feeble, and must, before its law was discovered, seem very irregular, the experiment made no strong impression on the audience."

Oersted communicated his discovery in the same year, in the *Annals of Philosophy*, in the paper *Experimenta* circa *effectum conflictus electrici in acum magneticam* ("Experiments on the effects of electric conflict on a magnetic needle"). Describing the interaction between the wire and the compass, he called the "effect which takes place in this conductor and in the surrounding space—*conflict of electricity*",[13] a denomination that was not adopted by other authors. Oersted visualized the magnetic interaction occupying the space around the wire:[14] "It is sufficiently evident from the preceding facts that the electric conflict is not confined to the conductor, but dispersed pretty widely in the circumjacent space."

This work had immediate repercussions all over the world, and Oersted became a well-known figure in scientific circles. In 1824, he founded the Danish Society for the Promotion of Natural Science, and in 1829, he became director of the Polytechnic Institute in Copenhagen.

What Hans Christian Oersted had observed was the fact that an electric current creates (or has associated with it) a magnetic field in its vicinity. This

---

[11] This account of how the discovery happened is reinforced by corrections pasted by Oersted onto the manuscript of the text sent for publication in "*The Edinburgh Encyclopaedia*" (1830). Within a text containing very few corrections, Oersted had written three versions of his account of the discovery before settling on a final and stronger fourth version, where he stated that the observed results were "strictly connected with his [Oersted's] other ideas." S. L. Altmann [39].

[12] H. C. Oersted, "Thermo-electricity," in *Edinburgh Encyclopaedia* (1830), XVIII, 573-589, quoted by L. Pearce Williams [40].

[13] H. C. Oersted, "*Annals of Philosophy*," 1820, in S. Sambursky [41].

[14] H. C. Oersted, "*Annals of Philosophy*," 1820, quoted by S. Sambursky [42].

field has an intensity that is directly proportional to the intensity of the current and points along a direction perpendicular to the current flow. The field is everywhere tangential to circles in the planes perpendicular to the conductor of electricity.

Oersted's experiment proved for the first time, beyond any doubt, the connection of electricity and magnetism. This was also the first experiment of conversion of electric energy (stored in the pile) into mechanical energy—the motion of the compass needle.

In his last work, *The Soul in Nature*, which Hans Christian Oersted left unfinished when he died in 1851, he proclaimed his faith in the unity of the universe: "Spirit and nature are one, viewed under two different aspects. Thus, we cease to wonder at their harmony" [37].

Magnetism and Electric Currents

Hans Christian Oersted's discovery produced a wave of excitement among researchers all over Europe. On 4 September 1820, the French physicist Dominique Arago (1786–1853) announced it at a meeting of the Académie des Sciences in Paris. Arago initially did not believe the result, but later, on the 11th of September, was able to repeat Oersted's experiment. Stimulated by the announcement of Oersted's discovery, another French physicist, André Marie Ampère (1775–1836) (Fig. 4.5), started a series of experiments on the relation of electric currents and magnetic fields, reporting his first observations within a few weeks of Arago's announcement. Ampère's findings would put his name in the pantheon of the founders of the science of magnetism.

Ampère was born in 1775, in Poleymieux-au-Mont-d'Or, near Lyon, the son of a wealthy merchant, and his life was marked by the social upheaval France was going through; in 1793, with the capture of Lyon by the Republican army, his father was guillotined.

Ampère was a man of wide interests: in 1819, he taught philosophy, and in the next year, he became assistant professor of astronomy at the University of Paris. In 1824, he was appointed professor of experimental physics at the Collège de France. His scientific work was also wide-ranging and included a classification of the chemical elements, published in 1816, that partly anticipated the work of the Russian Dmitri Mendeleev (1834–1907) on the periodic table of elements. His most important contribution, however, had to do with the relations between electricity and magnetism.

Based on his experiments on the motion of the compass needle under the action of electric currents, he invented an instrument to measure the intensity

Fig. 4.5 André Marie Ampère (1775–1836), French physicist who made important contributions to the study of electricity and magnetism

of these currents—the galvanometer, the ancestor of the present-day standard bench instruments, the voltmeters, and ammeters.

Between 5 and 17 September 1820, Ampère repeated Oersted's experiment and found that the compass needle points precisely in a direction perpendicular to the current-carrying wire when he compensated for the effect of the Earth's magnetic field. He also discovered that the current flows through every component of the electric circuit, including the voltaic pile [43]. Ampère presented a series of papers in 1820 at the Académie des Sciences on the subject of the interplay of electricity and magnetism. On 18 and 25 September 1820, he reported his observations and put forward his hypothesis on the relationship between the two classes of phenomena. He observed that parallel wires through which electricity flowed interacted with one another: if the current flowed in the same direction, the wires were attracted; if the directions were opposite, they were repelled. He demonstrated this finding on 9 October at the Académie.

There is some difficulty in establishing the correct chronology of Ampère's discoveries for two reasons: in the first place, because the text of the Memoirs of the Académie does not coincide with the content of the oral presentation, and second, due to the somewhat confused form of his communications [44]. This lack of clarity was recognized by Ampère himself and was also witnessed by Hans Christian Oersted when the two scientists met in Paris in 1823.

Oersted reported in harsh words his impression of Ampère: "He is dreadfully confused and is equally unskilled as an experimenter and as a debater".[15]

Throughout his research, Ampère always aimed at the mathematical formulation that described the physical phenomena connecting electric charges and magnetism. He derived a mathematical expression relating the forces applied between two wires that carried electric currents to the intensity of these currents and to the distance between the wires. Ampère's most important work, *Mémoire sur la Théorie Mathématique des Phénomènes Électrodynamiques Uniquement Déduite de l'Expérience* ("Note on the Mathematical Theory of Electrodynamic Phenomena Deduced Solely from Experiment"), was published in 1827.

Ampère showed that a coil made of copper wire, if free to turn, behaved in the same way as a compass needle. This confirmed the ideas that had occurred to him in the first days of his investigation [43], and led him to propose the bold hypothesis that the magnetism of the lodestone, or of the needle of the magnetic compass, arose from electric currents flowing inside matter: "…from which it follows that a magnet should be considered an assemblage of electric currents, which all flow in planes perpendicular to its axis, directed in such a way that the southern pole of the magnet, which is turned toward the north, is to the right of these currents while always to the left of a current placed outside the magnet, and which faces it in a parallel direction".[16]

The existence of this relationship of magnetism of matter with electric currents meant that the connection between electricity and magnetism was even more fundamental than the production of magnetic fields by electric currents in the Oersted experiment had demonstrated. The magnetism of the lodestone itself and of any magnetized body had a contribution from the electric currents that circulated within matter.

However, one paradox remained: in a magnet, the electric current required to explain its magnetism would heat it, but magnets were not known to be warmer than other objects! In fact, this objection against the first results of Ampère was pointed out very early by another French physicist, Augustin-Jean Fresnel (1788–1827) [47]. The solution to this puzzle was given by Ampère, at the end of 1820 and in 1821 (following a suggestion of the same Fresnel), when he explained that the currents responsible for the magnetism of matter were microscopic ("molecular"), rather than macroscopic, as the usual currents through conductors. He imagined a magnetic material

[15] Hans Christian Oersted, *Scientific Papers*, ed. K. Mayer, Copenhagen, 1920, p. cxiv, quoted by L. Pearce Williams [45].

[16] A. M. Ampère, "Recueil d'observations électrodynamiques" (1822), in S. Sambursky [46].

composed of molecules, and in each molecule, a coil where an electric current circulated. This idea anticipated the model for the atom proposed by the Danish physicist Niels Bohr (1885–1962) in 1913, which assumed that the positive atomic nucleus is surrounded by electrons that circulate around it. This atomic or molecular current would not produce the heating effects expected from ordinary, macroscopic currents.

The same explanation was imagined by Ampère to apply to the Earth's magnetic field: this field would arise from electrical currents flowing along the Equator; this was the first attempt to discuss the Earth's magnetism that had a sound scientific basis.

For his work on electricity and magnetism, Ampère was honored in the naming of the unit of electric current: the ampere, defined as the electric current transported when a charge of 1,602,176,634 $\times$ 10^{-19} coulombs flows through the conductor in 1 s. The ampere is one of the seven basic units of the International System of Units (SI), in use today through the Convention of the Meter established in 1960 and signed by most countries. The law that relates the magnetic field to the currents that produce it is also named after Ampère; it is one of the fundamental laws of electromagnetism (see Chap. 5).

The far-reaching consequences of Oersted's discovery of the generation of magnetic fields by currents pushed many investigators to search for the inverse phenomenon: electric effects produced by magnetic fields. Although such an inverse effect was expected, the efforts to observe it were mostly frustrated; in the words of the English physicist Michael Faraday, "…still it appeared very extraordinary, that as every electric current was accompanied by a corresponding intensity of magnetic action at right angles to the current, good conductors of electricity, when placed within the sphere of this action, should not have any current induced through them, or some sensible effect produced equivalent in force to such a current" [48].

And he concludes: "These consequences, the hope of obtaining electricity from ordinary magnetism, have stimulated me at various times to investigate experimentally the inductive effect of electric currents.

Michael Faraday (Fig. 4.6) was born in Newington (now Southwark, London), England, in 1791. His father, a blacksmith, had moved with his family to look for work in London. Faraday had very little formal education and lived through material hardship; he worked as an errand boy and, at the age of 14, entered a bookbinder shop as an apprentice. He then used every chance he had to read about almost every subject; his interest in science was awakened when he read the article "Electricity" in a volume of the *Encyclopaedia Britannica* that he was rebinding [49]. In 1810, he started to attend lectures

Fig. 4.6 Michael Faraday (1791–1867), the English scientist who made fundamental discoveries on electricity and magnetism; he also invented the electric motor and the dynamo

on physics and chemistry at the City Philosophical Society; in one of these lectures, Faraday saw a voltaic pile for the first time.

One of the clients of the bookbinder gave Faraday tickets to the popular science lectures of Humphry Davy (1778–1829) at the Royal Institution. Davy was a professor of chemistry and was famous for his excellent public lectures. In October 1812, Davy was temporarily blinded by an explosion that had occurred in the laboratory and therefore needed someone to help him on the occasion. Faraday was accepted by Davy as an amanuensis. Faraday then showed Davy the carefully bound notes of his public lectures that he had attended, and when a position opened the next year, he was hired as an experimental assistant.

In the same year, Davy took Faraday with him as he visited France and Italy on a tour that lasted over 1 year and included many scientific institutions. Faraday then had the opportunity of meeting some of the most important European scientists in different countries. Returning to London, he devoted himself to his experimental studies, mostly in chemistry. In 1821, he started to investigate electromagnetic phenomena, following the stimulus of his friend Richard Phillips (1778–1851). In the same year, he published in the *Quarterly Journal of Science* a paper where he described how a wire turned around a magnet when electric current flowed through it.

Faraday worked at the Royal Institution for 20 years and became one of the greatest experimentalists of all time. In his view, the required qualities of the scientist (philosopher) are the following: "The philosopher should be a man [*sic*] willing to listen to every suggestion, but determined to judge for himself. He should not be biased by appearances; have no favorite hypothesis; be of no school; and in doctrine have no master. He should not be a respecter of persons, but of things. Truth should be his primary object. If to these qualities he added industry, he may indeed hope to talk within the veil of the temple of nature".[17] Throughout his long scientific career, most of Faraday's experiments were of an exploratory character. In contrast with the ultimate goal of André Marie Ampère, Faraday's aim was not a mathematical description of the electromagnetic phenomena [51]; he rather tried to observe and carefully scrutinize new phenomena, asking in each situation how nature would behave under the given experimental conditions.

Humphry Davy was a close friend of the English poet Samuel Taylor Coleridge (1772–1834), an enthusiastic adept of *Naturphilosophie*; Faraday may have been influenced in this way by these philosophical ideas [52]. Imbued with the belief in the unity of physical phenomena, Faraday made several attempts to observe the inverse of Oersted's discovery, i.e., the generation of electric currents from magnetic fields. He searched for this inductive effect over a period of 11 years. According to his diary, he was finally successful in his trials on 29 August 1831. On that day, he had wound two coils of copper wire around a wooden cylinder, each 100 ft long; the two coils were electrically insulated one from the other, and at the ends of one coil, he connected a battery. To the other coil, he connected an instrument to detect electric currents (a galvanometer). Nothing was detected at the galvanometer while the current flowed through the coil. Nevertheless, a "slight deflection" was noted, but only at the moment the battery was connected and also when it was disconnected. Faraday also tried the same experiment with two coils wound on an iron ring; the same results were observed, and the intensity of the effect was much enhanced by the presence of the iron core.

In October of the same year, he made a very simple experiment that was closer to the realization of the inverse Oersted experiment: moving a magnet in and out through a hollow cylinder on which a coil had been wound, he obtained an electric current flowing through the coil. This current was detected by the deflection of a magnetic needle produced whenever the

[17] D. K. C. MacDonald, *Faraday, Maxwell, and Kelvin*, Anchor Books, New York, 1964, quoted by G. L. Verschuur [50].

magnet was moved relative to the coil. This effect was equally observed by moving the coil relative to the magnet.

The phenomena observed in this whole series of experiments can be explained in terms of induced electromotive force (emf) or voltage in the coil connected to the current detector (either a compass needle or a galvanometer). In the experiment with two coils, the primary coil, connected to the battery, produced a magnetic field at the other (secondary) coil. The time variation of this field induced a voltage in the secondary coil. The same applies to the use of the moving magnet to vary the field, instead of the primary coil. The faster the rate of change of the magnetic field (or magnetic flux) passing through the secondary coil, the larger the inductive effect, i.e., the emf, or voltage. This proved that such time-dependent magnetic fields created an electric field that acted on the electric charges, producing a flow of current. The mathematical expression of this fact is known today as Faraday's Law and is one of the fundamental equations of electromagnetism (see Chap. 6).

The experiment with the iron ring is comparable to the experiment with two coils, except that in this case, the magnetic field is augmented by the iron magnetization: the secondary coil samples both the magnetic field due to the first coil and the field due to the magnetization induced in the iron.

Faraday had proved that electric effects could be produced with magnets, but only if the corresponding magnetic fields were varying with time; this is why static experiments had not produced any detectable effect up to that point.

Faraday recognized at once the importance of these findings. He reported them to the Royal Society 1 month after his observation and got all the credit for the discovery of this new phenomenon, which became known as electromagnetic induction. In a well-known episode, Queen Victoria (1837–1901) asked Faraday about the utility of induction; "What is the use of a newborn child?" he retorted [53]. The twentieth century Czech poet and scientist Miroslav Holub (1923–1998) described the scene in these words [54]:

> When the Queen, over the
> magnetic lines of force
> on Faraday's rough table, asked
> And what use is it?
> Faraday replied,
> gazing lower than her
> lace collar:
> And what use, Ma'am, is a child?

Other researchers had stumbled upon the phenomenon of electromagnetic induction; the American physicist Joseph Henry (1797–1878) had noticed in 1829 a spark as he switched off power feeding a coil wound around another coil around an iron core. Henry would report his findings only in 1832 in the *American Journal of Science*, long after the publication of Faraday's paper; he would regret his delay in the submission of his results for his whole life [55]. Ampère, too, had seen the same effect even earlier, in 1822, but did not investigate it in detail [56]. A curious story is that of the Swiss physicist Jean-Daniel Colladon (1802–1893), who, in 1825, took the extra precaution of mounting his measuring instrument, a galvanometer, in another room to avoid the effect of electrical disturbances as he switched the current on and off. Each time he reached the other room to check the galvanometer, however, the effect of the inductance had already decayed, and he missed it altogether! [57]

The assembly of two coils wound on an iron ring is the first electrical transformer; feeding the primary coil with an oscillating voltage (usually referred to as AC or alternating current) produces another AC voltage on the second coil, and the voltage in the latter coil is proportional to the ratio of the number of turns in the two coils. Electromotive forces (emf) or voltages can then be stepped up or down, depending on this ratio.

If Oersted had found the principle that would lead to the electric motor, Faraday's discovery of the phenomenon of electromagnetic induction revealed the principle of the dynamo, or electric generator. If one adds the contribution of James Clerk Maxwell (Chap. 6), one has the scientific foundations that would spur the great leap in industrialization in the second half of the nineteenth century. In this period, for the first time in history, the accomplishment of technical advances required at least some familiarity with the developments that had taken place in the field of the pure sciences [58].

References

1. Planck, M. (1970). The unity of the physical world-picture. In S. Toulmin (Ed.), *Physical reality, essays on twentieth-century physics* (pp. 1–27). Harper & Row, p. 3.
2. Pliny the Elder. (1991). *Natural history: A selection, book XXXVII, no. 48* (p. 370). Penguin Books.
3. Roller, D. H. D. (1959). *The De Magnete of William Gilbert* (p. 22). Menno Hertzberger.
4. Plato. (1978). *Meno, [80a], Great books* (Vol. 7, p. 179). Encyclopaedia Britannica.

5. Aristotle. (1978). *History of animals, 620b [18], Great books* (Vol. 9, p. 146). Encyclopaedia Britannica.
6. Sarton, G. (1927). *Introduction to the history of science* (Vol. I). Robert E. Krieger Publishing Company. Reprinted 1975, p. 241.
7. Wu, C. H. (1984). Electric fish and the discovery of animal electricity. *American Scientist, 72*, 598–607, p. 601.
8. Wolf, A. (1962). *History of science, technology, and philosophy in the 16th and 17th century* (Vol. 1, 2nd ed., p. 303). George Allen & Unwin.
9. Bauer, E. (1958). L'Electricité et le Magnetisme au XVIIIe siècle. In R. Taton (Ed.), *Histoire Générale des Sciences* (Vol. II, p. 523). Presses Universitaires de France.
10. Bauer, E. (1958). L'Electricité et le Magnetisme au XVIIIe siècle. In R. Taton (Ed.), *Histoire Générale des Sciences* (Vol. II, p. 525). Presses Universitaires de France.
11. Cohen, B. (1957). Benjamin Franklin. In *Lives in sciences, a Scientific American book*. Simon and Schuster. (reprinted from Scientific American, August 1948), p. 117.
12. Wolf, A. (1962). *History of science, technology, and philosophy in the 18th century* (Vol. 1, 2nd ed., p. 230). George Allen & Unwin.
13. Mottelay, P. F. (1975). *Bibliographical history of electricity and magnetism* (p. 152). Arno Press.
14. Wolf, A. (1962). *History of science, technology, and philosophy in the 18th century* (Vol. 1, 2nd ed., p. 234). George Allen & Unwin.
15. Bauer, E. (1958). L'Electricité et le Magnetisme au XVIIIe siècle. In R. Taton (Ed.), *Histoire Générale des Sciences* (Vol. II, p. 521). Presses Universitaires de France.
16. Wolf, A. (1962). *History of science, technology and philosophy in the 18th century* (Vol. 1, 2nd ed., p. 224). George Allen & Unwin.
17. Wolf, A. (1962). *History of science, technology, and philosophy in the 18th century* (Vol. 1, 2nd ed., p. 222). George Allen & Unwin.
18. Bossi, L. (1994). L'Âme Electrique. In J. Clair (Ed.), *L'Âme au Corps, Arts et Sciences 1793–1993* (pp. 160–180). Galimard, p. 163.
19. Brown, T. M. (1980). Galvani. In C. C. Gillispie (Ed.), *Dictionary of scientific biographies* (p. 268). Charles Scribner's Sons.
20. Heilbron, J. L. (1980). Volta. In C. C. Gillispie (Ed.), *Dictionary of scientific biographies* (p. 76). Charles Scribner's Sons.
21. Bossi, L. (1994). L'Âme Electrique. In J. Clair (Ed.), *L'Âme au Corps, Arts et Sciences 1793–1993* (pp. 160–180). Galimard, p. 169.
22. Heilbron, J. L. (1980). Volta. In C. C. Gillispie (Ed.), *Dictionary of scientific biographies* (p. 69). Charles Scribner's Sons.
23. Wolf, A. (1962). *History of science, technology, and philosophy in the 18th century* (Vol. 1, 2nd ed., p. 256). George Allen & Unwin.

24. Heilbron, J. L. (1980). Volta. In C. C. Gillispie (Ed.), *Dictionary of scientific biographies* (p. 79). Charles Scribner's Sons.
25. Sambursky, S. (1974). *Physical thought: From the presocratics to the quantum physicists*. Pica Press.
26. Guthrie, W. K. C. (1967). *The Greek Philosophers—From Thales to Aristotle* (p. 37). Methuen & Co.
27. Guthrie, W. K. C. (1967). *The Greek philosophers—From Thales to Aristotle* (p. 36). Methuen & Co.
28. Barrow, J. D. (1991). *Theories of everything* (p. 17). Clarendon Press.
29. Adams, S. (1999). A theory of everything. *New Scientist, 161*(2174), A1.
30. Guthrie, W. K. C. (1967). *A history of Greek philosophy* (Vol. I, p. 70). Cambridge University Press.
31. Shanahan, T. (1989). Kant, Naturphilosophie, and Oersted's Discovery of electromagnetism: A reassessment. *Studies in History and Philosophy of Science, 20*, 287–305, p. 294.
32. Dibner, B. (1962). *Oersted and the discovery of electromagnetism* (p. 64). Blaisdell Publishing Company.
33. Shanahan, T. (1989). Kant, Naturphilosophie, and Oersted's discovery of electromagnetism: A reassessment. *Studies in History and Philosophy of Science, 20*, 287–305.
34. Shanahan, T. (1989). Kant, Naturphilosophie, and Oersted's discovery of electromagnetism: A reassessment. *Studies in History and Philosophy of Science, 20*, 287–305, p. 298.
35. Thuiller, P. (1990). De la Philosophie à l'Électromagnetisme: le Cas Oersted. *Recherche, 21*, 344–350.
36. Kline, M. (1985). *Mathematics and the search of knowledge* (p. 217). Oxford University Press.
37. Williams, L. P. (1980). Oersted. In C. C. Gillispie (Ed.), *Dictionary of scientific biographies* (p. 185). Charles Scribner's Sons.
38. Stauffer, R. C. (1953). Speculation and experiment in the background of Oersted's discovery of electromagnetism. *Isis, 48*, 33–50, p. 41.
39. Altmann, S. L. (1992). *Icons and symmetries* (p. 40). Oxford University Press.
40. Williams, L. P. (1980). Oersted. In C. C. Gillispie (Ed.), *Dictionary of scientific biographies*. Charles Scribner's Sons.
41. Sambursky, S. (1974). *Physical thought: From the presocratics to the quantum physicists* (p. 379). Pica Press.
42. Sambursky, S. (1974). *Physical thought: From the presocratics to the quantum physicists* (p. 380). Pica Press.
43. Williams, L. P. (1983). What were Ampère's earliest discoveries in electrodynamics? *Isis, 74*, 492–508, p. 507.
44. Williams, L. P. (1983). What were Ampère's earliest discoveries in electrodynamics? *Isis, 74*, 492–508.

45. Williams, L. P. (1983). What were Ampère's earliest discoveries in electrodynamics? *Isis, 74*, 492–508, p. 494.
46. Sambursky, S. (1974). *Physical thought: From the presocratics to the quantum physicists* (p. 385). Pica Press.
47. Hofmann, J. R. (1987). Ampère, electrodynamics and experimental evidence. *Osiris, 2nd series 3*, 45–76, p. 48.
48. Faraday, M. (1978). *Great books* (Vol. 45, p. 265). Encyclopaedia Britannica. (Original work published 1831).
49. Williams, L. P. (1980). Faraday. In C. C. Gillispie (Ed.), *Dictionary of scientific biographies* (p. 528). Charles Scribner's Sons.
50. Verschuur, G. L. (1993). *Hidden attraction: The history and mystery of magnetism* (p. 82). Oxford University Press.
51. Hofmann, J. R. (1987). Ampère, electrodynamics and experimental evidence. *Osiris, 2nd series 3*, 45–76, p. 76.
52. Williams, L. P. (1980). Faraday. In C. C. Gillispie (Ed.), *Dictionary of scientific biographies* (p. 530). Charles Scribner's Sons.
53. Forbes, R. J., & Dijksterhuis, E. J. (1963). *A history of science and technology* (Vol. 2, p. 456). Penguin Books.
54. Holub, M. (1995). Magnetism. *Poetry Review, 85*, 29.
55. Wilson, M. (1957). Joseph Henry. In *Lives of science, a Scientific American book* (p. 141). Simon and Schuster. (reprinted from Scientific American, July 1954) pg. 148.
56. Hofmann, J. R. (1987). Ampère, electrodynamics and experimental evidence. *Osiris, 2nd series 3*, 45–76, p. 68.
57. Lipson, H. S. (1968). *The great experiments in physics* (p. 103). Oliver & Boyd.
58. Hobsbawm, E. J. (1969). *Industry and empire, the Pelican economic history of Britain, Volume 3, from 1750 to the present day* (p. 173). Penguin Books.

Further Reading

Dibner, B. (1962). *Oersted and the discovery of electromagnetism*. Blaisdell Publishing Company.

Sambursky, S. (1974). *Physical thought: From the presocratics to the quantum physicists*. Pica Press.

Sarton, G. (1927). *Introduction to the history of science* (Vol. I). Robert E. Krieger Publishing Company, reprinted (1975).

Wolf, A. (1962). *History of science, technology, and philosophy in the 16th and 17th century* (Vol. 1, 2nd ed.). George Allen & Unwin.

5

"Acting Where It Is Not": Magnetism and Action at a Distance

I have no new discovery to bring before you this evening. I must ask you to go over very old ground, and to turn your attention to a question which has been raised again and again ever since men began to think. The question is that of the transmission of force.
James Clerk Maxwell, in a communication at the Royal Society, in the year 1854.

Summary The idea that all matter is constituted of small particles (atoms) was first proposed by Democritus (c. 460–c. 370 BC), a Greek thinker. This hypothesis was given a scientific justification much later, as the English researcher John Dalton (1766–1844) studied the proportion of the elements required to form a compound, e.g., the volumes of hydrogen and oxygen, to form water (H_2O). Only in the twentieth century did the New Zealander Ernest Rutherford (1871–1937) determine that atoms have a tiny nucleus, around which the electrons turn. The Danish physicist Niels Bohr (1885–1962) proposed the first model that describes the atoms. The German Max Planck (1858–1947) suggested that atoms absorbed and irradiated energy in packets, or "quanta." The German-American Albert Einstein (1879–1955) later advanced the idea that light and radio waves carried discrete packets of energy, or quanta. The ideas of Planck, Einstein, and others defined a new physics, subsequently to be known as quantum mechanics. In this conception, all matter—including the particles—presents wavelike properties. One of the triumphs of this new theory is that it explained the origin of magnetism, which rests on a quantum interaction ("exchange") between electrons.

A. P. Guimarães, *A Longstanding Attraction*, https://doi.org/10.1007/978-3-032-02006-2_5

Magnetic Attraction

The most amazing aspect of magnetism has always been the effect of one object on another without any visible material link between them. This may be observed with a magnet and a piece of iron or with two magnets. The interaction between two magnets can be simply described through the concept of poles (Chap. 4). In each magnet, there are two poles, North and South, somewhat point-like, that exert forces on the poles of the other magnet. With like poles, the force is repulsive, and with opposite poles, it is attractive. Therefore, if like poles of two magnets are nearer to each other than the opposite poles, the overall effect will be one of repulsion. However, if two magnets are placed far from one another, the forces between like poles and the forces between opposite poles will be approximately the same, and the final result will be neither attraction nor repulsion. In this condition, the forces applied on the two poles will, in general, have equal intensities and opposite directions, and their sum is zero; they may, however, produce a torque. This is the case of the force applied by the Earth's magnetic field on the needle of a compass: the total force is zero (i.e., the needle is not attracted to either Earth pole), but there is a tendency to turn that aligns the needle in the North–South direction. We may then say that in a uniform magnetic field, such as that acting on the needle, there is no overall force, and the magnetic attraction that one associates with the action of the magnet is in fact only observed in nonuniform fields, as in the immediate neighborhood of a magnet.

Using two long bar magnets, one can measure the forces that one pole of one magnet exerts on another pole of the other magnet. By measuring the force and varying the separation between them, it can be determined how the force varies with distance. This was done by the French physicist Charles-Augustin de Coulomb (1736–1806), who published his results in 1785 in the *Mémoires de l'Académie Royale des Sciences*; he found that this force is proportional to the inverse of the square of the distance. When the distance doubled, the magnetic force was reduced to one-fourth of its original strength. Coulomb also found the same inverse square dependence for the force between electrically charged objects.

As we bring a piece of iron close to the magnet, it behaves as if it has acquired magnetism. The iron is "magnetized," that is, it temporarily becomes another magnet, and the force of mutual attraction between the iron and the magnet is of the same type as that between two magnets, as described above.

The phenomena of magnetic attraction and repulsion were explained by most Greek philosophers in antiquity in terms of contiguous action, i.e., action through an intervening medium: the lodestone acted on this medium, which in turn carried the influence of the stone to the piece of iron. The idea of direct action at a distance was, in general, rejected.

However, with the beginnings of modern science, in the sixteenth and seventeenth centuries, the concept of action at a distance gained increasing acceptance, and magnetic attraction became the paradigm of such action. Finally, the idea of action at a distance reached its culmination with Isaac Newton's theory of gravitation.

It was thought in antiquity in the Western world that two bodies could only affect one another if they were in direct physical contact, and if this was not the case, something material between them transmitted the influence of one to the other. For instance, for Lucretius (Titus Lucretius Carus), a Latin philosopher and poet who lived in the first century BC, the presence of air was instrumental in bringing about the attraction between the magnet and iron: "a stream flows from this stone that pushes the air, creating a vacuum; this attracts the iron" [1]. The great philosopher Plato, who lived in Greece from c. 428 to c. 348 BC, had expressed a similar view in the Timaeus [2].

In antiquity, the notion of an intervening medium was reinforced by the Stoics, who invoked a *pneuma*, from the Greek for "air" or "breath." An all-pervading mixture of air and fire that had elastic properties, the *pneuma* was able to transmit forces at a distance. An example of an effect at a distance through an intervening medium is given in the writing of Theon of Smyrna (c. AD 100), describing the phenomenon of resonating cords: "Strings are in resonance with each other—if on a string instrument one of them is struck, then the other, by some kinship and sympathy, sounds in accord".[1]

Other authors often saw magnetic attraction as arising from a flow of vapors, or "effluvia," or a stream of corpuscles. This is found in the view of Lucretius, quoted above, and in that of the Neo-Platonists. In late antiquity, the Neoplatonists attributed effects at a distance to the propagation of corpuscles that could push or impart a pulling force, the latter in the case of magnetic attraction [4]. Along the same lines, and much earlier, the Greek philosopher Empedocles of Acragas (c. 490–c. 435 BC), who lived in the fifth century before our era, regarded the action of the magnet on iron (as reported by his contemporary Alexander) as arising from effluences, or vapors, emitted through the pores of both magnet and iron.

The same idea of effluvia survives to the modern age, as shown, for example, in the seventeenth century writings of the great French philosopher and mathematician René Descartes (1596–1650). Descartes, in his *Principles of*

[1] Theon of Smyrna, Expos. Rer. Math. 50, 22, in S. Sambursky [3].

Philosophy (*Principiorum Philosophiae*) of 1644, attributes the magnetic attraction of iron to the magnet to "threaded particles" (*particulae striatae*, in the Latin original) that "expel air between them, making them approach each other".[2] Two currents of particles flowed through the magnet, one in each direction.

Isaac Newton (1642–1727) himself, praised above all as the creator of the theory of Universal Gravitation, hesitated in accepting the possibility of action at a distance. To suppose "that one body may act upon another, at a distance through a vacuum, without the mediation of anything else, (…) is to me so great an absurdity, that I believe no man who has in philosophical matters a competent faculty for thinking, can ever fall into it," he wrote[3] in the year 1692, in a letter to the scholar and priest Richard Bentley (1662–1742).[4]

In the East, the ideas of action at a distance were more widely accepted, since they fit better into the framework of Chinese philosophy. This philosophy generally favored reciprocal relations, rather than relationships in a single direction. The difference between the Chinese and the Western view in mechanics would mean an emphasis on actions at a distance over actions of mechanical shock [8].

The Arab philosopher Averroës (Ibn Rushd) (1126–1198) explained the action of the magnet through a mechanism he called "multiplication of the species"; this meant that the lodestone modified the air near it, which in turn modified the air further away until this influence arrived at the piece of iron, which then was acted upon by this "virtue."

The knowledge of magnetic phenomena resulted in the establishment of magnetism as the paradigm of action between non-contiguous bodies. This tendency was reinforced with the work of William Gilbert (1544–1603), through his celebrated book *De Magnete*, published in 1600 (see Chap. 3); for Gilbert, magnetism was the determining force that held together the solar system.

Subtle Matter: The Ether

The idea that the presence of a medium was required to carry interactions between faraway objects is very old. The Greek Stoic philosopher Chrysippus (c. 280–207 BC), from Soli, who lived in the third century before Christ,

---

[2] R. Descartes, "*Principes de Philosophie*," quoted by R. Lenoble [5].

[3] Isaac Newton, quoted in Arthur Koestler [6].

[4] According to Karl Popper, Newton condemned here, by anticipation, all his followers [7].

described vision as mediated by a tension in the air between the eye and the object. The original name given to the go-between medium was *pneuma*. In the modern age, from the seventeenth century, many writers invoked an ether (or aether) from the Greek *aither*, which originally meant air or fire [9]. This would be a medium responsible for carrying the attraction between the planets or, in general, responsible for the influence between noncontiguous objects affected by gravitational or magnetic forces. For the Greeks after Aristotle, and probably for many thinkers before him [10], the ether filled the upper heavens, and this was the stuff stars and planets were made of [11]; above earth, water, air, and fire, the ether was the "fifth element" that constituted the universe.

William Gilbert did his experiments on magnetism at the end of the sixteenth century. Gilbert and other thinkers, like René Descartes, thought that some fluid matter existing around a magnet had vortices that explained the magnetic attraction. Descartes used the same explanation for the gravitational attraction [12]: "let us assume that the material of the heaven where the planets are circulates ceaselessly, like a whirlpool with the sun as its centre, and that the parts which are near the sun move more quickly than those which are a certain distance from it." Therefore, bodies did not exert forces on each other across empty space; instead, bodies were acted upon through whirlpools in the medium that filled space.

The requirement of the existence of this medium filling space was also postulated to explain the propagation of light. The Dutchman Christiaan Huygens (1629–1695) was one of the proponents of this point of view. Within the mechanistic framework, Huygens, knowing that sound waves propagate more rapidly in less compressible fluids, and also that the speed of light was so immense, concluded that "the particles of the ether to be of a substance as nearly approaching to perfect hardness and possessing a springiness as prompt as we chose" [13]. Other thinkers held that the propagation of light required something material, like Descartes, who did not believe in the existence of vacuum: all space was filled with particles—space was a *plenum*. And through these particles, "the action of light is communicated" [14].

Many authors shared this notion that light required a medium as a support for its propagation. Newton also expressed the same view in his treatise on Optics (1704) [15].

Toward the end of the nineteenth century, James Clerk Maxwell (1831–1879) (Fig. 5.1) discussed the propagation of light, initially in terms of stresses in a medium filling space. He later moved away from a purely mechanical picture, refining his ideas in such a way that his ether was regarded by the proponents of fluid models as "too ethereal" [16].

Fig. 5.1 James Clerk Maxwell (1831–1879), Scottish physicist who first formulated the theory of electromagnetism

The idea of an ether lived until it was shaken by an experiment on the propagation of light along different directions relative to the motion of the Earth, performed by the German-born American physicist Albert Abraham Michelson (1852–1931) in 1881 and later repeated with the assistance of the American chemist Edward Williams Morley (1838–1923) several times until 1929. The experiment consisted of the comparison of the time taken by two perpendicular light rays, traveling along paths of the same length in the north–south and east–west directions, made [17] with very precise techniques that used the interference[5] of light. This famous experiment demonstrated that light propagated with the same velocity along the direction of motion of the Earth around the Sun, or perpendicular to this direction. This was not the expected result if the Earth moved relative to an ether filling interplanetary space where light propagated.

This disagreement had an important impact on the ideas of how light could travel in a vacuum. The final blow that made the idea of ether completely superfluous was provided by Albert Einstein (1879–1955) with his special theory of relativity, published in 1905. No support was necessary for the propagation of light; it was accepted that light (and, as we shall see below, all the other electromagnetic radiations) could travel through empty space.

[5] Interference is the effect arising from the superposition of two waves. When two identical waves moving in the same direction meet, and the crests and troughs of them coincide, the effect is that the two waves add up, and the total amplitude is doubled. If, on the other hand, the crests of one wave coincide with the troughs of the other, the effect is a cancellation. In the case of a light wave incident on a screen, cancellation creates a dark band and superposition, a light band.

Fig. 5.2 Isaac Newton (1642–1727), English physicist and mathematician, one of the greatest scientists of all time

The Apple and the Moon

As we have already mentioned, the idea of action at a distance was fully established with the triumph of the Newtonian gravitation theory. The first ideas of this kind were expressed in more or less precise form by many thinkers, e.g., by the astronomer Johannes Kepler (1571–1630), who understood that two stones, depending on their masses, "would come together, after the manner of magnetic bodies".[6] Also, in the writings of the English philosopher and man of letters, Francis Bacon (1561–1626): "Many powers act and take effect only by actual touch, as in the percussion of bodies (...) Other powers act at a distance, though it be very small, of which but few have as yet been noted; although there may be more than men suspect" [19].

It was with Newton, however, that the action-at-a-distance point of view reached its culmination. Isaac Newton (Fig. 5.2) was born on Christmas Day 1642, near the village of Colsterworth, in Lincolnshire, England. He was a premature baby who grew to become a solitary youth and entered the University of Cambridge at the age of 18.

In 1665, Cambridge students were sent home to avoid the Great Plague, an epidemic that killed, in London alone, some 70,000 people, which represented about one-seventh of London's population. Newton, who had graduated in the same year, returned to the family farm in Woolsthorpe and lived

[6] Johannes Kepler, "*Introduction to Astronomia Nova*" (1609), quoted in Arthur Koestler [18].

there for almost 2 years, in a remarkably creative period known as his *anni mirabilis*, the wonder years in the history of science. Many of his discoveries, including calculus, the binomial theorem, the decomposition of white light into the colors of the rainbow, and above all, the foundations of the gravitation theory, were made in those years. Fifty years later, he described his achievements in that period with the following words: "In the beginning of the year 1665 I found the Method of approximating series & the Rule for reducing any dignity [degree] of any Binomial into such a series. The same year in May I found the method of Tangents of Gregory & Slusius, & in November had the direct method of fluxions [differential calculus] & the next year in January had the Theory of Colours & in May following I had entrance into the inverse method of fluxions [integral calculus]. And the same year I began to think of gravity extending to the orb of the Moon & (having found out how to estimate the force with which [a] globe revolving within a sphere presses the surface of the sphere) where Kepler's rule of the periodical times of the Planets being in sesquialterate [i.e., as the power 3/2] proportion of their distances from the center of their Orbs, I deduced that the forces which keep the Planets in their Orbs must [be] reciprocally as the squares of the distances from the centers about which they revolve: & thereby compared the force requisite to keep the Moon in her Orb with the force of gravity at the surface of the earth, & found them answer [agree] pretty nearly".[7]

This incredible list of discoveries marks a breathtaking period in the intellectual history of humankind.

The name of Isaac Newton is usually associated with the theory of gravitation through the well-known anecdote according to which Newton had a flash of insight as he watched an apple falling from a tree in his garden. His notes, however, reveal that Newton arrived at the concept of gravitation through careful analysis, not as the result of a sudden revelation. Revelation or not, the authenticity of an episode with Newton that involved the fall of an apple in Lincolnshire in the year 1666 is confirmed by at least three independent contemporaries [20].

If gravity reached a branch of an apple tree, being responsible for the fall of the fruit, why wouldn't it reach the moon, in this case, affecting its motion around the Earth? If the Earth did not exist, the moon would move along a straight line. One could think that at each instant that the moon deviated from this straight line in its almost circular (elliptical) orbit, it was "falling" to the Earth, in analogy with the apple.

[7] Quotation from Newton's recollections, Westfall, op. cit., p. 39.

This was a bold step: to relate the matter-of-fact notion of weight, the tendency of bodies to fall down, to phenomena occurring in the sky. The laws of nature that ruled the sublunar world might extend their kingdom to the realm of the heavens, a concept in complete opposition to the classical Aristotelian view of a world separated into supralunar and sublunar spheres.

The path that led Newton to his discovery of universal gravitation started from the understanding that the Moon had weight, like the ordinary Earth objects; then he assumed that the same force acted between the Sun and planets, and planets and satellites; finally, that this type of interaction was a universal principle, applicable to every object in the universe [21].

Although Newton was reluctant to accept the idea of action at a distance (see the above quotation from a letter written in 1692 to Richard Bentley), he came to adopt it as the foundation of his theory of universal gravitation. His profound interest in alchemy[8] made it possible for him to accept this type of action, an element common in the occult sciences [22].

His view on magnetic attraction was more ambiguous. In the 31st Query in the Opticks, he puts together gravity, magnetism, and electricity to affirm that 'the small Particles of Bodies [have] certain Powers, Virtues, or Forces, by which they act at a distance (…) upon one another for producing a great Part of the Phaenomena of Nature'.[9]

Although recognized in his lifetime as a great scientist, Newton displayed a mean side of his character in many controversies and priority disputes in which he was involved, especially with the German philosopher and mathematician Gottfried Wilhelm Leibniz (1646–1716) on the paternity of the calculus, and with the English physicist Robert Hooke (1635–1703) on the discovery of the inverse-square law of attraction.

Newton's most important published work was the book *Philosophiae Naturalis Principia Mathematica* (Mathematical Principles of Natural Philosophy), known as the *Principia*, which appeared in 1687 and required the support and dedication of the astronomer Edmund Halley to be published. The second treatise that summarized Newton's researches was the *Opticks*, published in 1704, at a time when he was the president of the Royal Society. The *Opticks* was written in the form of numbered Queries (16 in the first edition), embodying the results of his discoveries since his time as an undergraduate at Cambridge. In a version of the *Opticks* planned in the early

[8] "We have every reason to suppose that, had Newton *not* been steeped in alchemical and other magic learning, he would never have proposed forces of attraction and repulsion between bodies as the major feature of his physical system" [22].

[9] Isaac Newton, *Opticks, or a treatise of the reflections, refractions, inflections & colours of light*, New York, 1952, based on the 4th edition, London, 1730, pp. 375-6, quoted by R. W. Home [23].

1690s, Newton had included in Book IV a demonstration of the existence of forces that act at a distance; he now left this out, fearing his critics. In the second English edition, he added queries 17–24 that dealt with the existence of the ether.

The force of gravity was finally described on a scientific basis within Newtonian mechanics. The idea of action at a distance was applicable to gravitational attraction, as well as to forces between electric charges and to magnetic attraction and repulsion.

At the end of his life, Newton was acclaimed in England as the greatest scientific authority and founder of a new world order. At the same time, Newton's ideas were not accepted by his contemporaries in continental Europe, where the principle of action at a distance was regarded as archaic, representing a relapse into Aristotelian physics [24]. Mechanical models, which attributed attraction between the heavenly bodies to whirlpools in the ether, were still favored on the continent.

Newton finished the last edition of the *Principia* in 1726 and died in Kensington the following year; on his tomb, an inscription honored him in no mean terms: "Mortals rejoice that there has existed such and so great an ornament to the human race!"

After some time, the winds started to change on the continent: with the growth of empiricism and the success of Newtonian physics and astronomy, the acceptance of Newton's ideas was finally established.[10]

Lines of Force Fill Space

Two centuries after the *Principia*, the English physicist Michael Faraday (1791–1867) wrote: "How the magnetic force is transferred through bodies or through space we know not" [25]. Following this statement, he speculated if that interaction was of the same type as the forces between electric charges, whether or not it required the existence of an ether. In his study of electricity, Faraday initially thought in terms of action through contiguous atoms: he was inspired by the way one charged body induces charges of opposite sign on the nearest surface of a neighboring body, which may then produce the same effect on a third body. But with time, his thoughts tended gradually in the direction of abandoning the idea of action at a distance.

Faraday studied the patterns of the iron filings around the poles of a magnet. The filings form chains that draw curved lines converging to the two

[10] M. B. Hesse, op. cit., p. 166.

opposite poles. The points of convergence coincide with regions of more intense magnetic field. This led Faraday to represent the magnetic field with lines of force [or field lines] drawn in space, such that their direction at each point of space was the same as the direction of the magnetic forces on a piece of iron at that point (or more precisely, as the direction of a compass needle at that point). His concept of lines of force evolved from a mere form of describing the fields to "physical" lines that mediated the interactions and had all the "reality." *This represented a shift in emphasis that would lead to the concept of field.* "Faraday relegated the particle to the background and enthroned in its stead lines of force throughout space," in the words of one of his biographers [26]. And moreover, Faraday, in his *Experimental Researches,* regarded the medium or space around it as essential as the magnet itself.[11]

The stage was set for the birth of the modern concept of field, an idea that had as an early ancestor the *pneuma* and was also somewhat related to the concept of ether.

The Triumph of the Fields

Before Faraday, the idea of action at a distance had also arisen in the context of the understanding of the properties of solid matter, among them the impenetrability of matter. The Jesuit Roger Boscovich (1711–1787) produced in the eighteenth century an approach to the problem of the constitution of matter that brought to the fore the interactions between atoms. For him, the atoms themselves were point-like, and their presence was detected only through their power to attract or to repel. This was also the view of his contemporary, the English clergyman and scientist Joseph Priestley (1733–1804), who argued that solid matter appeared so because of the force of repulsion felt at a distance from the objects.[12]

James Clerk Maxwell (1831–1879), the outstanding Scottish physicist whose words were quoted in the opening of this chapter, was concerned with the problem of action at a distance; in the same work, in continuation, he wrote [28]: "We see that two bodies at a distance from each other exert a mutual influence on each other's motion. Does this mutual action depend on the existence of some third thing, some medium of communication,

[11] M. Faraday, "Experimental Researches," quoted by Herbert Kondo, p. 138.

[12] "To conclude that resistance, on which alone our opinion concerning the solidity or impenetrability of matter is founded, is never occasioned by solid matter, but by something of a very different nature, viz. a power of repulsion always acting at a real and in general an assignable distance from what we call the body itself" [27].

occupying the space between the bodies, or do the bodies act on each other immediately, without the intervention of anything else?"

James Clerk Maxwell created his theoretical synthesis upon the foundations set by the experiments of Michael Faraday. Maxwell was born in Edinburgh in 1831 and revealed his aptitude for mathematics very early, winning at the age of 14 a prize with an essay on ovals, which was read before the Royal Society of Edinburgh by a university professor. He joined the University of Edinburgh when he was 16 and, in 1850, enrolled at the University of Cambridge.

After obtaining his degree, Maxwell stayed for 2 years at Trinity College, Cambridge, where he read the Experimental Researches of Faraday. For some years, he studied the kinetic theory of gases, a subject to which he made significant contributions, using a simple picture of molecules as hard particles, like billiard balls. He created the well-known image of "Maxwell's demon," an imaginary being that could sort fast molecules from slow molecules, thereby separating a hot portion (fast particles) from a cold portion (slow particles) of a gas, an enterprise forbidden, under normal circumstances, by the Second Law of Thermodynamics.

In 1860, he became a professor at King's College in London and, for 5 years, dedicated himself to studies of electricity and magnetism. He took over the concept of lines of force from Faraday and tried to formulate a theory that would describe them within a sound mathematical framework. He then produced a model based on the dynamics of fluids, making an analogy between pressure in a hypothetical fluid and electric potential. He showed, in the paper "On Faraday's Lines of Force," that there was a perfect analogy between the shape of the lines of force of the electrostatic field and the lines of flow of an incompressible fluid.

His definitive theory was published in the year 1864 in "A Dynamical Theory of the Electromagnetic Field"; the mechanical images were then reduced to a minimum, and the electromagnetic phenomena were described in terms of fields present on a substrate that had adequate mechanical properties—the ether.

In 1846, in a paper entitled "Thoughts on Ray-Vibrations," Faraday had speculated on the possibility of electromagnetic oscillations: "The view which I am so bold as to put forth, considers (…) radiation as a high species of vibrations in the lines of force which are known to connect particles and also masses of matter together. It endeavors to dismiss the ether, but not the vibrations".[13]

[13] Quoted by Herbert Kondo [29].

In 1855, the German physicists Wilhelm Eduard Weber (1804–1891) and Rudolf Kohlrausch (1809–1858) measured the charge of a capacitor with two different experiments and derived from the result that the ratio of the values obtained in two different systems of units, called electromagnetic and electrostatic, was a number very close to the known velocity of light. This suggested that the electrical disturbances traveled at the speed of light. Maxwell made the daring assumption that light and electrical disturbances were one and the same thing—electromagnetic waves. "We can scarcely avoid the inference," he wrote [30].

The remarkable advances reached in the studies of Maxwell can be summarized in four equations, known to this day as "Maxwell's equations," which, for Electromagnetism, are analogous to the Ten Commandments for Christianity. They embody in quantitative form the description of every phenomenon studied by his predecessor, Michael Faraday; these equations connect the electric charges and electric currents to the values of the associated electric and magnetic fields. For static fields, two equations deal with the electric part, and the other two with the magnetic part. However, if the fields vary with time, the intensities of the electric and magnetic terms are related.

Maxwell's equations describe every electromagnetic phenomenon and even predict the possibility of existence and the form of propagation of electromagnetic waves; they constitute the pinnacle of classical physics.

A few years after Maxwell's death, the German physicist Heinrich Rudolf Hertz (1857–1894) detected radio waves and showed in his experiments in Karlsruhe, between 1885 and 1889, that they had many properties of light waves; they were reflected in the same way, propagated in a straight line, exhibited refraction, and so on. This confirmed experimentally Maxwell's identification of light as electromagnetic radiation. In Hertz's own words[14] "(…) this is true for light as such or any special sort of light: of the sun, of a candle, of a glowworm."

The difference between light waves and radio waves resides only in their different wavelengths (or frequencies, with which there is a one-to-one correspondence in a vacuum). Waves of light have a length of about 1 μm (0.001 mm), whereas radio waves have wavelengths above 1 mm; for example, FM radio and TV transmissions use waves about 3 m long, and AM radio hundreds of meters long. There are electromagnetic waves of much shorter wavelengths, such as X-rays (0.00001 μm) or gamma rays (one-millionth of a micrometer and below).

[14] H. Hertz, "Gesammelte Werke," vol. 1, 1895, quoted in S. Sambursky [31].

But what do we mean when we speak of an electromagnetic wave? It is instructive to make a parallel with a water wave. When a stone is thrown into a lake, it produces water waves that propagate from the point of fall; what we see is a disturbance that moves. A given water molecule on the surface of the lake will be set in motion as the water wave progresses past it. One may describe this wave as the oscillation in height that affects molecules further and further away from the center of the ripples. An electromagnetic wave also propagates in time, but it needs no material medium. As an electromagnetic wave passes through a given point in the vacuum, what oscillates at that point is the magnitude of the magnetic and electric fields.

The concept of field, as well as the idea of its physical reality, i.e., the claim that it was more than a mere mathematical construct, became a cornerstone of Maxwell's theory. This concept made a strong impression on the young Albert Einstein. He wrote in his Autobiographical Notes: "The most fascinating subject at the time that I was a student was Maxwell's theory. What made this theory appear revolutionary *was the transition from forces at a distance to fields as fundamental variables* [my emphasis APG]. The incorporation of optics into the theory of electromagnetism (...)—it was like a revelation" [32].

According to the field point of view, one electric charge does not exert a force on another distant charge, but instead, the electric field associated with the first charge acts on the second charge. The field is the agent, not the first charge. We usually say that an electric charge "creates" an electric field around it, but in fact, the charge and the field are interconnected and form a whole that cannot be separated.

Physicists give a general definition of a field as a continuous distribution in space of values of some physical quantity: one may, therefore, speak of a field of temperatures in a room, meaning that to each point in space, one may attribute a value to the temperature. Another example of a field is the ensemble of values of water velocities at different points within a stream of flowing water.

The concept of a field is useful when one wants to describe the effects of an electric charge or of a magnet in its immediate vicinity. In the neighborhood of an electric charge, for example, another charge will experience a force. The quantity obtained by measuring the force on the second charge and dividing it by the value of this charge is called the magnitude of the electric field. The ensemble of values of this quantity, one for each point in space, is also called the "electric field" of the first charge.

Albert Einstein and Leopold Infeld (1898–1968), in their *Evolution of Physics*, hailed the concept of field as "the most important invention since Newton's time." And they added, referring specifically to electric and magnetic fields: "It needed great scientific imagination to realize that it is not the charges or particles but the field in the space between the charges and the particles which is essential for the description of physical phenomena".[15] In Einstein's worldview, matter and fields had the same status.[16]

An interpretation of electromagnetism that opened the way to Einstein's view was that of the Dutch physicist Hendrik Antoon Lorentz (1853–1928). Lorentz tried to follow a middle course between the action-at-a-distance point of view and a "field" concept. He hypothesized that there was an immobile ether that carried electric and magnetic interactions. To be consistent with this program, he had to move away from some of Newton's postulates; specifically, if charges exerted forces on the ether, it did not apply forces on them, in opposition to Newton's law of Action and Reaction [35]. He proceeded to derive how the electric and magnetic fields would appear to observers both at rest and in motion relative to an absolute frame of reference (identified with the ether). The results he obtained were essentially the same as those later derived by Einstein in his special relativity theory, except that in Einstein's theory, it was shown that the assumption of an ether was not necessary.

Authors tend nowadays to emphasize the conceptual importance of the idea of field, either re-affirming its "reality" or regarding it as a fundamental mathematical tool for the description of electromagnetic interactions [36]. Some have seen the field, e.g., the electric field, as a property acquired by the space itself in the vicinity of a charge; others as a new entity contained in this space, the space itself remaining simply a geometric container.[17]

Fields produced by static electric charges and fields produced by magnets are instances of a more general entity—the electromagnetic field. We recall that electric charges in motion generate magnetic fields. The unity of electric and magnetic fields is more clearly perceived if we compare two situations. Suppose we have an electric charge stationary with respect to an observer; this observer detects the electric field due to the charge. If now the charge begins to move, its presence is gradually perceived by this observer also through its magnetic field. Another observer, on the other hand, travelling with the particle, would always observe only an electric effect.[18] In short, the magnetic and

[15] A. Einstein and L. Infeld, *The Evolution of Physics*, Cambridge University Press, 1938, quoted in [33].

[16] Or in the language of philosophy of science: "Einstein put the field ontologically on a par with matter" [34].

[17] In the words of the Italian physicist Giuliano Toraldo di Francia (1916–2011): "(..) it is also a metaphysical prejudice to allow empty space to have only *geometrical* properties and to discard the possibility that it may have also *physical* properties. It took nearly a century of theoretical and experimental work before physicists began to see this possibility" [37].

[18] An observer that observes a stationary electron detects in the neighborhood of the particle an electric field. If the particle is in motion relative to the observer, it is also detected a magnetic field. This change is described by the relativity theory (see, e.g., a discussion in Piccioni [38]).

electric parts of the electromagnetic field are manifestations of the same field—electromagnetic—for different observers. This arises from the relativity principle, as formulated by Einstein; although Einstein's relativity theory was created after the findings of Maxwell, Maxwell's equations are consistent with relativity, and in fact, they were at the root of Einstein's theorizing.

Electromagnetic waves are a general denomination for both radio and light waves; they travel in empty space with a velocity of about 300,000 km/s (186,000 mi/s). Today, physicists know many different fields besides the electromagnetic field and the gravitational field. For example, another field acts inside protons and neutrons, binding together the quarks that form these constituent particles of the atomic nucleus.

We have so far spoken of fields in classical terms: in physics, this means the terms previous to the advent of quantum mechanics, the new physics developed from the 1920s (Chap. 6). In the non-classical descriptions, in the context of quantum electrodynamics (QED), a special theory that incorporated the contributions of quantum mechanics and relativity, developed from the 1950s, the subatomic particles (electrons, muons, quarks, etc.) are related to the fields in the sense that they are concentrated bundles, or quanta, of different types of fields. The masses of these particles vary over a very wide range: the heaviest quark—the top quark—is more than 100,000 times heavier than the electron.

The types of interactions between these particles, called "forces," are also described by fields; for example, the electric and magnetic interactions involve quanta called photons.[19] These photons participate in the interaction of attraction or repulsion involving charges. Nuclear forces also have their corresponding quanta, in this case called gluons; there are eight types of gluons. All these particles in the interaction fields are virtual particles, which means that they are not directly observable. They are not static; instead, they are continuously being created and annihilated, and the interactions are the result of the exchange of these particles, e.g., between the charges, in the case of the electric field. This interaction field interacts with the other type of field, the matter field.

In the QED description of elementary particle physics, there are only fields, and this view embodies, in the words of the American Nobel Prize-winning particle physicist Steven Weinberg[20] (1933–2021), "(...) the central dogma of quantum field theory: *the essential reality is a set of fields* subject to the rules of

---

[19] A good description of the evolution of this picture of elementary particle physics is given by Davies and Brown [39].

[20] S. Weinberg, quoted by S. Y. Auyang [40].

special relativity and quantum mechanics;(...)."[21] The theory that describes these fields is called the Standard Model of particle physics.[22]

The revolution that started with Michael Faraday one and a half centuries ago led, in the framework of quantum field theory, to a new view of the world, a world of interactions between fields: matter fields and interaction fields. However, most users of magnets do not think in terms of such complex and abstract entities. Their mental picture is different; magnets still interact with iron at a distance, with the interaction somehow represented by the vivid image of the lines of field, imprinted into our minds since Faraday's time.

References

1. Lucretius. (1978). *The great books* (Vol. 12, p. 93). Encyclopaedia Britannica.
2. Plato. (1978). Timaeus. In *The great books* (Vol. 7, p. 471). Encyclopaedia Britannica.
3. Sambursky, S. (1962). *The physical world of late antiquity* (p. 101). Princeton University Press.
4. Sambursky, S. (1962). *The physical world of late antiquity* (p. 102). Princeton.
5. Lenoble, R. (1958). Le Magnetisme et l'Électricité. In R. Taton (Ed.), *Histoire Générale des Sciences* (Vol. II, pp. 333–334). Presses Universitaires deFrance.
6. Koestler, A. (1973). *The sleepwalkers* (p. 344). Penguin Books.
7. Popper, K. R. (1972). *Conjectures and refutations* (p. 107). Routledge and Kegan Paul.
8. Haudricourt, A., & Needham, J. (1966). La Science Chinoise Antique (Chapter V). In R. Taton (Ed.), *Histoire Générale des Sciences* (Vol. I, Part I, p. 198). Presses Universitaires de France.
9. Guthrie, W. K. C. (1967). *A history of Greek philosophy* (Vol. I, p. 271). Cambridge University Press.
10. Guthrie, W. K. C. (1967). *A history of Greek philosophy* (Vol. I, p. 136). Cambridge University Press.
11. Guthrie, W. K. C. (1967). *A history of Greek philosophy* (Vol. I, p. 466). Cambridge University Press.
12. Sambursky, S. (1974). *Physical thought: From the presocratics to the quantum physicists* (p. 248). Pica Press.

[21] "(...) to field physicists, the fields constitute the fundamental ontology. Spacetime is a structural property of the fields, not the other way around" [40].

[22] This theory predicts, to account for the masses of the subatomic particles, the existence of still another kind of field, the scalar fields. These fields differ from, e.g., the magnetic field, since the latter has a direction (and is therefore generally described by a vector) while scalar fields do not. The particles of the scalar fields are usually called Higgs particles.

13. Huygens, C. (1978). Treatise on light (Ch. I). In *The great books* (Vol. 34, p. 559). Encyclopaedia Britannica.
14. Descartes, R. (1971). Meteorology (1637). In M. P. Crosland (Ed.), *The science of matter* (p. 71). Penguin.
15. Newton, I. (1978). Optics. In *The great books* (Vol. 34, pp. 520–521). Encyclopaedia Britannica.
16. North, J. D. (1990). *The measure of the universe: A history of modern cosmology* (p. 29). Dover.
17. Shankland, R. S. (1964). The Michelson-Morley experiment. *Scientific American, 211*, 107.
18. Koestler, A. (1973). *The sleepwalkers* (p. 342). Penguin Books.
19. Bacon, F. (1978). Novum organum. BII. In *The great books* (Vol. 30, p. 176). Encyclopaedia Britannica.
20. Westfall, R. S. (1993). *The life of Isaac Newton* (p. 51). Cambridge University Press, 308.
21. Bernard Cohen, I. (1992). *The birth of a new physics* (p. 238). Penguin.
22. Henry, J. (1989). Newton, matter and magic. In J. Fauvel, R. Flood, M. Shortland, & R. Wilson (Eds.), *Let Newton be* (p. 144). Oxford Press.
23. Home, R. W. (1992). *Electricity and experimental physics in eighteenth-century Europe* (p. 256). Variorum.
24. Hesse, M. B. (1962). *Forces and fields: The concept of action at a distance in the history of physics* (p. 157). Greenwood Press.
25. Faraday, M. (1978). *The great books* (Vol. 45). Encyclopaedia Britannica, Twenty-eighth series, 1851, p. 759.
26. Kondo, H. (1957). Michael Faraday. In *Lives of science, a Scientific American book* (p. 127). Simon and Schuster. (reprinted from Scientific American, October 1953) p. 138.
27. Priestley, J. (1971). Disquisitions relating to matter and spirit (1777). In M. P. Crosland (Ed.), *The science of matter* (p. 116). Penguin.
28. Maxwell, J. C. (1965). Proceedings of the Royal Institution. In W. D. Niven (Ed.), *The scientific papers of James Clerk Maxwell* (Vol. VII, Part II, p. 311). Dover.
29. Kondo, H. (1957). Michael Faraday. In *Lives of science, a Scientific American book* (p. 127). Simon and Schuster. (reprinted from Scientific American, October 1953), p. 139.
30. Newman, J. R. (1957). James Clerk Maxwell. In *Lives of science, a Scientific American book* (p. 155). Simon and Schuster. (reprinted from Scientific American, June 1955), p. 172.
31. Sambursky, S. (1974). *Physical thought: From the presocratics to the quantum physicists* (p. 461). Pica Press.
32. Einstein, A. (1969). Autobiographical notes. In P. A. Schilpp (Ed.), *Albert Einstein: Philosopher-scientist* (Vol. I, p. 33). Cambridge University Press.
33. Crosland, M. P. (Ed.). (1971). *The science of matter* (p. 343). Penguin.

34. Nersessian, N. J. (1984). *Faraday to Einstein: Constructing meaning in scientific theories* (p. 35). Martinus Nijhoff Publishers.
35. Brown, L. M. (Ed.). (1993). *Renormalization from Lorentz to Landau (and beyond)* (p. 43). Springer-Verlag.
36. Antoniazzi, M., & Giuliani, G. (1997). Campi, onde e particelle nei manuali di elletromagnetismo: un esempio di stratificazione concettuale e ontologica. *Giornale di Fisica, 38*, 87.
37. Toraldo di Francia, G. (1976). *The investigation of the physical world.* Cambridge University Press.
38. Piccioni, R. G. (2007). Special relativity and magnetism in an introductory physics course. *The Physics Teacher, 45*, 152.
39. Davies, P. C., & Brown, J. (1992). *Superstrings: A theory of everything?* Cambridge University Press.
40. Auyang, S. Y. (1995). *How is quantum field possible?* (p. 150). Oxford University Press.

Further Reading

Cohen, I. B. (1992). *The birth of a new physics*. Penguin.
Davies, P. C., & Brown, J. (1992). *Superstrings: A theory of everything?* Cambridge University Press.
Koestler, A. (1973). *The sleepwalkers*. Penguin Books.
Sambursky, S. (1962). *The physical world of late antiquity*. Princeton University Press.
Westfall, R. S. (1993). *The life of Isaac Newton*. Cambridge University Press.

6

The Secrets of Matter: Like Flies in a Cathedral

For the first time a plausible story can be told concerning the ultimate magnetic particles, the essential nature of the atom of a ferromagnetic substance, the kind of forces which determine the properties of magnetic crystals (...).
(R. M. Bozorth, in The Physical Basis of Ferromagnetism *(1940). Bell Systems Technical Journal vol. 19 (1940) 1, p. 2, quoted by Stephen T. Keith and Pierre Quédec [1]).*

Summary How does the matter of the lodestone differ from other substances? The understanding of magnetism had to wait until the atomic constitution of matter was established, and the laws that rule the atomic and subatomic world were known in the first decades of the twentieth century. The idea of atoms has a long history; one of the earliest proposers was Democritus (c. 460–c. 370 BC). In the eighteenth century, Antoine-Laurent Lavoisier (1743–1794) characterized the idea of an element. John Dalton (1766–1844), from the proportions in weight of elements that entered a chemical reaction, established that atoms are the smallest units of matter. Dmitri Mendeleev (1834–1907) drew a chart with 63 chemical elements, the "periodic table." Joseph John Thomson (1856–1940) identified negative particles, later known as electrons. Max Planck (1858–1947) proposed in 1900 the existence of the quantum. Albert Einstein (1879–1955) quantified the radiation. Ernest Rutherford (1871–1937) discovered in 1909 that the atom is mostly empty space. Niels Bohr (1885–1962) created in 1912 the first quantum description of the atom. Louis de Broglie (1892–1987) established that all matter has wave properties. Erwin Schrödinger (1887–1961) and Paul

A. P. Guimarães, *A Longstanding Attraction*, https://doi.org/10.1007/978-3-032-02006-2_6

Dirac (1902–1984) formulated equations to describe quantum systems. Dirac characterizes the post-1925 years as the Golden Age of Physics.

The central concept of atomism is that every single object is formed of small, immutable particles that cannot be divided (*atomon*—in Greek, that cannot be cut, indivisible) [2]. The Greek philosopher Simplicius of Cilicia, writing in the fifth century AD in *Commentary on the Physics*, stated[1]: "(Democritus of Abdera (c. 460—c. 370 BC) posited the full and the void as first principles, one of which he called being and the other non-being; for he posits the atoms as matter for the things that exist and generates everything else by their differences." To form the material bodies, the atoms stick to one another, since some of them are concave, some convex, some have hooks, and so on.[2]

The idea of the discrete nature of matter is found in Lucretius, a Roman poet and philosopher of the first century BC, a supporter and advocate of atomism. In his *De Rerum Natura* ("On the Nature of Things"), Lucretius explained a series of natural phenomena on an atomistic basis. Thus, he argued that the smallness of the atoms explains why the senses cannot detect the amount of metal that is rubbed each day from a worn ring or from an iron plowshare [5]. The atoms would confer characteristic properties to different substances in the world around us: for example, things that are pleasant to the senses would be formed of round and smooth atoms.

Some Indian philosophers who lived several centuries before our era were also atomists. They had the Vedic texts Nyāya and Vaiśeṣika as authoritative [6].

In the modern era, the seventeenth century philosopher René Descartes also considered that matter was formed of particles; for example, water was formed of "long, smooth, and slippery" particles [7]. However, in his *Principles of Philosophy* (1644), he argues that since extension is an attribute of everything that exists (and therefore occupies space), there can be no indivisible particles, or atoms. Furthermore, matter could, in principle, be continuously divided since the opposite would mean that God had deprived Himself of this power [8].

Thus, the question of continuity versus discreteness of matter was addressed by a variety of thinkers since the time of Democritus, with differing approaches and emphasis. Finally, by the end of the nineteenth century, the accumulated knowledge on electricity and electromagnetic phenomena, as well as the

---

[1] Simplicius, Commentary on the Physics, quoted by J. Barnes [3].

[2] Simplicius, Commentary on the Heavens, quoted by J. Barnes [4].

growing corpus of chemical knowledge, favored a new turn in the discussion of this age-old problem.

"The Real Facts of Nature"

The early pre-Socratic philosophers were the first thinkers to attempt to describe the whole of the universe as formed of a single principle, or element (Chap. 2). As we have seen, this element, for Thales of Miletus (640–545 BC), was water. Among other thinkers who suggested primary elements, there were Anaximenes (588–524 BC), who suggested air, and Heraclitus, fire. Empedocles of Acragas, or Agrigento, in Sicily (c. 490–c. 435 BC), posited that the multitude of substances in the world around us stems from combinations of the four elements: earth, water, air, and fire. This concept was later also adopted by Aristotle, who saw the qualities of the different substances reflecting the proportion of these four elements in their composition. This, in essence, remained the dominant view for two thousand years.

Robert Boyle (1627–1691), an English scientist, wrote in *The Sceptical Chymist*, in 1661, against Aristotle's theory of the four elements, arguing that all matter was formed of primary corpuscles, in his words [9], 'divided into little particles of several sizes and shapes variously moved.' One hundred and twenty years later, in the second half of the eighteenth century, the Frenchman Antoine-Laurent Lavoisier (1743–1794), one of the founders of modern chemistry, elaborated on the concept of element, giving an operational definition: a substance that could not be analyzed or divided by chemical means. Lavoisier, who also innovated in the use of quantitative techniques in chemistry, established in his *Traité Élémentaire de Chimie* ('Elementary Treatise of Chemistry') of 1789 that there existed 33 elements.

"An inquiry into the relative weights of the ultimate particles of bodies is a subject, as far as I know, entirely new;" announced the English chemist and physicist John Dalton (1766–1844) in a communication [10] presented at the Literary and Philosophical Society of Manchester in 1805. And he added, 'I have lately been prosecuting this inquiry with remarkable success.'

The work of John Dalton gave, in the first years of the nineteenth century, a scientific basis for the belief that atoms were the smallest units of matter. An atom was the smallest portion of a given element that had the same chemical properties as the bulk substance; the smallest portion of a compound was a molecule. Dalton arrived at these conclusions from the study of the proportions in weight of the different elements that entered a chemical reaction; for example, hydrogen and oxygen always participated in the proportion of 2:1 in the chemical reaction that gave rise to water.

In 1869, a Russian professor of chemistry, Dmitri Ivanovich Mendeleev (1834–1907), presented a chart containing 63 of the then-known chemical elements, organized in increasing atomic weights. This table revealed similarities in the chemical properties of the elements and had empty spaces that were associated with elements to be discovered. The fact that this chart gave emphasis to several properties of the elements that occurred repeatedly along the series made it known as the "periodic table." This table went through many revisions, including a rearrangement according to another characteristic of the elements, their atomic number. This form of presentation was a major step in the study of the elements, and the scientific importance of the periodic table has been compared to that of Darwin's theory of evolution [11].

Still in the nineteenth century, studies conducted by Michael Faraday on the chemical decomposition produced by the flow of electricity and by another English physicist, William Crookes (1832–1919), on electric discharges through a gas at low pressure suggested that there existed negative electric charges.

Joseph John Thomson (1856–1940), a British physicist, wrote: "(…) I can see no escape from the conclusion that they are charges of negative electricity carried by particles of matter. The question next arises, What are these particles?" [12] These particles of negative charge studied by J. J. Thomson were later called 'electrons' and were shown by him to be about 2000 times lighter than the lightest atom.

The electron was the first subatomic particle to be identified, and its discovery represents an enormous advance in the understanding of the constitution of matter. We may also say that this fact inaugurated the discipline of elementary particle physics.

These advances occurred in parallel with progress in the comprehension of the structure of matter, more specifically on the understanding of the fact that matter is made of atoms and on the nature and constitution of these ultimate building blocks. Perhaps the most important experiment in the quest for understanding the structure of the atom was that performed under the supervision of Ernest Rutherford (1871–1937), a New Zealander physicist who moved to England in 1895. Rutherford had spent some years in Montreal, where he studied the physical properties of alpha and beta particles, determined the first radioactive half-life, and conceived the radioactive transmutation theory in 1902 (with the English chemist Frederick Soddy (1877–1956)). He returned to England in 1906, this time converting the University of Manchester into a world center for research in physics and practically creating a discipline that would subsequently be designated "nuclear physics." For his remarkable discoveries made in Canada on the radioactive decay of nuclei, Rutherford was awarded the Nobel Prize for chemistry in 1906.

The celebrated experiment was performed in 1909 by Rutherford's assistant, Hans Geiger (1882–1945), and a student, Ernest Marsden (1889–1970). It consisted of bombarding a thin foil of gold with alpha particles, which are energetic nuclei of the element helium, emitted in the decay of some radioactive nuclei. It revealed that the majority of alpha particles went right through the atoms, as expected, but a very small proportion of them were reflected or strongly deviated. "It was quite the most incredible event that has ever happened to me in my life," Rutherford later remarked [13]. "It was almost as incredible as if you fired a 15-inch shell at a piece of tissue paper and it came back and hit you."

Rutherford correctly interpreted this surprising result as meaning that the atom is almost completely void, with a very small and massive nucleus, with a radius 100,000 times smaller than the atomic radius. The electrons, in Rutherford's vivid image,[3] whizzed through the empty space around the nucleus, "like a few flies in a cathedral."

Ernest Rutherford was convinced of the primacy of experimental interrogation of the secrets of nature. Speaking[4] of the theoretical physicists, who use as their tools the mathematical description of the physical world, he once said, "They play games with their symbols," and he added, "but we turn out the real facts of Nature." It has been pointed out that Rutherford was a type of diversifying mind, interested in exploring the multiplicity of natural facts, in opposition to other scientists, like Einstein, who emphasized in their work the search for a unifying picture of the world.[5]

In 1911, Rutherford met the young Danish physicist Niels Bohr (1885–1962), who had graduated from the University of Copenhagen, where he also obtained his PhD, and was then doing postdoctoral work in Cambridge. Bohr was captured by Rutherford's charisma and decided to come to Manchester to work with him.

Niels Bohr returned to Denmark in 1912 and, in the next year, published three papers where he proposed the first theory of the atom that involved quantum concepts. He had to postulate that electrons circled the nucleus in stationary orbits, in a picture similar to the solar system. However, in direct

[3] E. Rutherford, quoted by Victor Guillemin [14].

[4] Ernest Rutherford, quoted by Freeman Dyson, [15].

[5] The American physicist of British origin, John Freeman Dyson (1923–2020) sees a parallel in these two attitudes, which could be described as "Baconian" and "Cartesian," with the cities of (coincidently) Manchester, the first industrial city, and Athens, the first academic city Freeman Dyson [16].

opposition to the principles of classical physics, the electronic orbits in his model could only have discrete radii. Classically, any value of orbital radius would be allowed, and the electron would occupy a continuum of energy states, rather than a set of discrete energy values. In addition, according to classical electromagnetism, a charged particle would eventually spiral into the nucleus, and therefore, the atom would be unstable.

A significant development that contributed to the knowledge of the structure of matter was the discovery of a mysterious radiation emitted by uranium salts by the French physicist Antoine Henri Becquerel (1852–1908). The phenomenon of radioactivity, as it became known later, was accidentally stumbled upon as Becquerel noticed in 1896 that a photographic film was exposed in the proximity of the uranium compounds.

The Polish physicist Marie Sklodowska Curie (1867–1934) and her husband, Pierre Curie (1859–1906), working at the University of Paris, found in 1898 that samples containing the element thorium were also radioactive. After 10 years of investigation by the Paris group and by Rutherford (1871–1937), then at McGill University in Montreal, it became clear that radioactive atoms emitted three types of rays: alpha rays, beta rays, and gamma rays. This would later be identified, respectively, as helium nuclei, electrons, and energetic photons. In 1894, in an often quoted remark, the German-born American physicist Albert Michelson (1852–1931), speaking of the *fin-de-siècle* perspectives of physics, concluded: "It seems probable that most of the grand underlying principles have been firmly established and that further advances are to be sought chiefly in the rigorous applications of these principles to all the phenomena which come under our notice" [17]. In a glaring rebuttal of Michelson's appraisal, in an interval of a few years, in the 1890s, some very important discoveries heralded a profound change in physics, with repercussions that would have a very broad reach, affecting every sphere of human activity. These included the discovery of X-rays (1895), radioactivity (1896), and the electron (1897).

Despite the momentous scientific advances[6] at the end of the nineteenth century, the actual constitution of matter was still unknown. The existence of atoms as fundamental building blocks of all matter and the comprehension of their internal structure would be definitely established only in the first decades of the twentieth century.

[6] In some cases, the potentialities of these advances were immediately presumed. For example, a contemporary writing on the observation of X-rays described it as "(…) a discovery so strange that its importance cannot yet be measured, its utility be even prophesied, or its ultimate effect upon long-established scientific beliefs be even vaguely foretold." H. J. W. Dam, *McClure's Magazine*, April 1896, quoted in Muldawer, L. [18].

The Quantum World

On 14 December 1900, at a meeting of the German Physical Society, Max Planck (1858–1947) (Fig. 6.1), then professor of theoretical physics at the University of Berlin, presented his novel ideas on heat and radiation. He had been studying the frequency spectrum of the radiation emitted by the walls of a furnace, a property that is independent of the material that constitutes these walls. The experimental results that he was analyzing had been obtained at the laboratories of the Physikalisch-Technische Reichsanstalt, the imperial bureau of standards in Berlin, some years before. This frequency spectrum, or distribution, is related, for example, to the fact that the color of a heated object changes as the temperature increases from a dull red to a bright yellow.

In his work, Planck considered matter composed of atoms, which acted as oscillators (like radio receivers and transmitters) that absorbed and emitted radiation. The novel assumption he made was that the energy of these oscillators assumed discrete values, or energy "quanta" (the plural of the word *quantum*, neuter of the Latin word for how much). This proposal allowed him to explain the frequency distribution of the radiation emitted by heated bodies. It was a revolutionary idea, representing a direct challenge to the prevailing theories, part of the edifice of what came to be known as classical physics, in which the energy of a physical system may assume any value, varying continuously. Although Planck saw this assumption of discreteness, or quantization, as the only way to explain the experimental facts, he was not very comfortable

Fig. 6.1 The German physicist Max Planck (1858–1947), founder of the quantum theory

with it and many years later described [19] "the whole procedure as an act of despair."

Max Karl Ernst Ludwig Planck was born in Kiel, Germany, in 1858. He obtained his doctorate in Munich in 1879 on thermodynamics and became a full professor at the University of Berlin in 1892. Besides his qualities as a scientist, he was an excellent musician and, in his youth, considered the possibility of following an artistic career; he even composed an operetta and, in later years, played in a trio that included Albert Einstein [20].

Planck was for many decades the most important name in the German scientific establishment, "the voice of scientific research,' [21] as rector of the Berlin University from 1913, secretary of the Berlin Academy, and the most influential figure of the Physical Society, and from 1930 president of the Kaiser-Wilhelm-Gesellschaft (Kaiser Wilhelm Society). The latter scientific research organization had been founded in 1911 and, in 1948, was renamed the Max Planck Society.

One of the consequences of the original 1900 work of Max Planck was that it gave support to the idea of the atomic constitution of matter, at that time still an object of controversy. For the Swedish chemist Svante Arrhenius (1859–1927) [22], this was "the most important offspring" of Planck's proposal.

In the 5 years following Planck's paper, practically no further progress was made in the direction that he had inaugurated. In 1905 and 1906, however, the then-unknown young patent officer in Bern, Albert Einstein (Fig. 6.2), applied the same approach to treat two other physical problems. The first was

Fig. 6.2 Albert Einstein (1879–1955), a German-American physicist, is celebrated for his extraordinary scientific contributions that include the theory of relativity and the theory of the photoelectric effect

the so-called photoelectric effect, a phenomenon that consists of the appearance of a positive electric charge on some metals, produced by the incidence of light. This originates in the fact that electrons are emitted from the surface of these metals. However, the fact that their energy is directly related to the frequency of the light, but not to its intensity, could not be explained by classical physics. Einstein's proposal took Planck's hypothesis one stage further, assuming that the energy of the light, or electromagnetic radiation in general, was also carried in discrete packets, or quanta. This idea was not at first generally accepted, but would be credited in the end as the correct explanation for the effect.

The other significant extension of the new theory, also proposed by Einstein, was the "quantization" of the thermal vibrations of the atoms in a solid; according to this hypothesis, as a solid was heated, the vibrations of the atoms could have only discrete values of energy. With this suggestion, the relation between the amount of heat absorbed by a body and the increase in its temperature, involving a quantity that physicists call 'specific heat,' could be explained for a wide range of temperatures.

Albert Einstein was born in Ulm, Germany, in 1879 to a Jewish middle-class family. In his childhood, his family moved to Italy after his father's business failed. Later, he settled in Switzerland, where he studied to be a physics teacher at the Swiss Federal Polytechnical School (ETH) in Zurich. After he graduated, he took up a job at the Patent Office in Bern in 1902. He subsequently occupied different academic positions, remaining from 1914 to 1932 as a professor of physics at the University of Berlin. Einstein wrote the 1905 papers in Bern, practically isolated from other physicists; he once mentioned that the first time he met a professional physicist was when he was thirty.

Einstein's contributions to physics include the special relativity theory of 1905, the theory of Brownian motion, the general relativity theory of 1916, and many essential results in the branch of physics called statistical mechanics, which applies statistical methods to different physical phenomena. The general relativity theory dealt with the gravitational field and predicted that light rays would interact with this field. The observational confirmation of Einstein's prediction was achieved in 1919 as light from the stars was shown to deviate as it passed near the Sun. The success of the theory drew the attention of the world public opinion and made Einstein instantly famous.

While in Berlin, Einstein witnessed the growing power of the Nazi party and felt the asphyxiating climate created by the mounting anti-Semitic campaign that included a denunciation of "Jewish science" as unworthy and false. Under such pressure, he left Germany in 1932, finally taking up residence in the United States and joining the Institute of Advanced Studies in Princeton.

There, he was destined to play a fundamental role in history when he signed a letter to President Franklin Delano Roosevelt in 1939 advising that the Allies should undertake a project of tapping the energy from nuclei, since there were signals that Germany had already taken this path. This letter helped to set in motion a huge scientific, industrial, and military machine that would lead to the first nuclear bomb test in the Alamogordo desert on July 16, 1945.[7] The ensuing atomic bomb attack on Hiroshima and Nagasaki filled Einstein with grief, shocked his humanitarian conscience, and reinforced his pacifist militancy for the rest of his life.

Einstein's direct relation with magnetism, besides the recollection of the impression made by a compass in his childhood (see the opening quotation of this book), and a school essay written when he was only sixteen,[8] also includes his only experimental work. This was done in 1915 in the *Physikalisch Technische Reichsanstalt* and resulted in the discovery of the so-called Einstein-de Haas Effect. In this work, he tried to find an experimental proof of the electric currents postulated by Ampère and determined that the currents were real and that they imparted a torque to a rod of magnetized metal [25].

Albert Einstein became the best-known scientist of the twentieth century, and Time magazine chose him in the year 2000 as the most important person of that century.[9] He died in 1955, at the age of 76.

The proposals of Planck, Einstein, and others would form the basis of a new physics later to be baptized as quantum mechanics. This name was used for the first time in print[10] in 1914 by Arnold Eucken (1884–1952), editor of the German edition of the annals of the second Solvay Congress, held in Brussels in 1913.

The existence of the light quanta, known as photons, was confirmed in experiments made by the American physicist Arthur Holly Compton (1892–1962) and published in 1922. Compton's experiment showed that photons that had collided with electrons lost some of their energy and, therefore, changed their frequency, in agreement with the hypothesis that each quantum of light transported an amount of energy proportional to the

[7] Another letter sent much later, this time urging the American government not to use the bomb against Japan, was found unopened on Roosevelt's desk on the day of his death. Whitrow [23].

[8] A. Einstein, 'On the Investigation of the State of the Ether in the Magnetic Field', 1895, mentioned by Lewis Pyenson [24].

[9] *Time Magazine*, New York, December 31, 1999.

[10] A. Eucken, Anhang Die Theorie der Strahlung und Quanten, Knapp, Halle, 1914, p. 373, through Helmut Rechenberg [26].

frequency of the radiation. This interpretation of the experiment, however, was established only after some years of controversy [27].

In the years 1920–1922, Bohr discussed, based on quantum principles [28], the systematic variation of the properties of the elements of the periodic table. He succeeded in explaining how the chemical and physical properties of the elements varied along the table. The atoms of the elements, starting with hydrogen, with one electron, have increasing numbers of electrons; he assumed that as each electron was added, occupying a new state, the preceding electrons were not disturbed, remaining in their original states.

The French physicist Louis de Broglie (1892–1987), as part of his doctoral thesis in Paris, published in 1923 and 1924, the hypothesis that ordinary matter exhibited wavelike behavior. A particle behaves as a wave with a wavelength inversely proportional to its mass. Everyday objects, on the human scale, also have undulatory behavior that we do not detect because their intrinsic wavelength is too small. This daring idea had an experimental confirmation a few years later when Clinton Joseph Davisson (1881–1958) and Lester Halbert Germer (1896–1971) in the United States reported in 1927 that electrons presented wave behavior and that their wavelength agreed with de Broglie's prediction. These authors had demonstrated that a beam of electrons produced an 'interference' pattern, a succession of light and dark bands on a screen, very similar to that observed with X-rays, known to be electromagnetic waves.

This puzzling wave-particle duality is a very fundamental aspect of nature, which can be detected when one observes photons, electrons, or other particles on the atomic and subatomic scale. Each type of experiment brings to the fore either the particle or the wave character of these entities. According to J. J. Thomson,[11] it can be said that in this dual behavior, as in a struggle "between a tiger and a shark, each is supreme in his own element but helpless in that of the other."

The mathematical description of the wave properties of matter was accomplished by another founding father of quantum mechanics, the Austrian physicist Erwin Schrödinger (1887–1961), after he had read in 1925 de Broglie's thesis. In a seminar presented in the next year in Zurich, where he was a professor of physics, he announced: [30] "My colleague Debye [the Dutch physicist Peter Debye (1884–1966)] suggested that one should have a wave equation; well, I have found one!" Schrödinger's wave equation was applied to a multitude of physical problems, yielding solutions that widened the range of systems whose behavior is determined by the quantum character

[11] J. J. Thomson, *Structure of Light*, Cambridge, 1925, p. 15, quoted by B. R. Wheaton [29].

of matter. This included, above all, the physics of the atom and the subatomic particles.

The description and understanding of the simplest atom, the atom of hydrogen, was achieved in 1926 by Schrödinger. He was frustrated, however, when the next more complex atom, the atom of helium, failed to yield to his equation. Paul Adrian Maurice Dirac (1902–1984), an English physicist and another member of the team of founders of the new physics, commented to Schrödinger that this was not important; what really mattered was the aesthetic value of his theory. In a later account of his celebrated remark [31], Dirac maintained that "(…) it is more important to have beauty in one's equations than to have them fit experiment." Beauty was also more important than the principle of simplicity, or the economy of elements, or parameters, a paramount requirement or criterion in a scientific theory known as Ockham's razor,[12] a name given in honor of the English Franciscan and logician William of Ockham (c. 1284–1349).

The period from 1925 onwards was characterized by Dirac as the Golden Age of Physics [33]. He gave, in the years 1928–1931, a significant contribution to this age when he generalized Schrödinger's work to make the wave equation compatible with Einstein's theory of relativity. With this modification, there emerges another attribute of the electron, the "spin," analogous, in terms of classical mechanics, to an intrinsic spinning momentum. This had been demonstrated by the work of two Dutch physicists, Samuel Goudsmit (1902–1978) and George Uhlenbeck (1900–1988), in 1925. The spin is an essential concept in the understanding of magnetism, as will be discussed in the next section.

One of the most remarkable results that came out of the quantum picture is the possibility, in an experiment where two particles are emitted, as the quantum state of one of them is measured, of instantaneously determining the state of the other, irrespective of the distance between them. This is true even when the information from the first particle does not have time to reach the second particle, in an apparent violation of the cherished principle of causality. This experiment was not performed in Einstein's time, but in any case, he did not accept its predicted outcome. He thought[13] it would mean

[12] "The research worker, in his [sic] efforts to express the fundamental laws of Nature in mathematical form, should strive mainly for mathematical beauty. He should still take simplicity into consideration in a subordinate way to beauty. It often happens that the requirements of simplicity and beauty are the same, but where they clash the latter must take precedence." P. A. M. Dirac, 'The relation between mathematics and physics,' *Proc. Roy. Soc. (Edinburgh)* 59 (1938/1939) 122–9 (Feb. 25, 1939) (James Scott Prize Lecture), quoted by Helge Kragh [32].

[13] *The Born-Einstein Letters*, Irene Born, trans., Walker, New York, p. 158, quoted by N. David Mermin [34].

the result of "spooky actions at a distance." This strange property of nature-often called "non locality,"nevertheless, has found experimental verification by the French physicist Alain Aspect (1947–) at the University of Paris in 1982.

This effect arises from a special relationship between states of particles called entanglement, which, according[14] to Schrödinger, is '*the* characteristic trait of quantum mechanics.' Due to the property of entanglement, knowing the state of a pair of particles does not mean that the state of each individual particle is known, i.e., in some way, the particles are 'inseparable.' It forms the basis of the physical realization of "teleportation,' [36] which can be performed experimentally by passing the state of one particle to another, as demonstrated in the laboratory in 1997.

Quantum mechanics has been extremely successful in the description and in making accurate predictions of phenomena on the subatomic scale. Furthermore, mastering the quantum principles and applying them to the knowledge of the physics of solids has allowed the design and large-scale production of electronic components and devices. It has been estimated [37] that some 30% of the Gross National Product (GNP) of the United States depends on innovations that result from the knowledge of quantum mechanics.

Despite these successes, quantum mechanics has been controversial in its interpretation since its very first years. Einstein and Bohr, for example, held a protracted debate during their lifetimes since they held conflicting points of view on this interpretation. The mathematical equations and the values of physical quantities that are derived in quantum mechanics are not questioned, but how can we understand the counterintuitive concepts of this remarkable theory? What is the meaning of the fundamental wave function that enters Schrödinger's equation?[15] One author of several important contributions to quantum physics, the American physicist Richard Feynman (1918–1988), stated bluntly, '[I] think I can safely say that nobody understands quantum mechanics.'[16]

[14] E. Schrödinger, 'Discussion of the Probability Relations Between Separated Systems,' *Proceedings of the Cambridge Philosophical Society*, vol. 31 (1935) pp. 555–563, quoted by Jeffrey Bub [35].

[15] Erich Hückel (1896–1980), a young physicist at Zurich when Schrödinger presented his wave function, expressed his puzzlement in verse:

"Erwin with his psi can do
Calculations quite a few
But one thing has not been seen
Just what does psi really mean?" (Felix Bloch [38]).

[16] R. Feynman, "*The Character of Physical Law*," 1965, p. 129, quoted by J. T. Cushing [39].

Every revolutionary scientific theory stimulates the discussion of the fundamentals of science. Quantum mechanics, more than any other theory, was conducive to a reassessment of the ideas of causality and determinism, objectivity and subjectivity. The growth of quantum mechanics and the widening acceptance of the validity of its description of atomic and subatomic phenomena stimulated the discussion of the philosophical implications of quantum ideas. Not only philosophers participated in this debate, but also many scientists, mostly physicists, took up the challenge of interpreting the consequences of the new physics. "The physicist cannot simply surrender to the philosopher the critical contemplation of the theoretical foundations," argued Albert Einstein. Moreover[17]: "for, he himself knows best, and feels more surely where the shoe pinches."

The triumph of quantum mechanical ideas led to a widespread questioning of causality and determinism. Causality is essentially the basic assumption that natural phenomena have causes that precede them. The modern concept of causality was already expressed [41] by the English philosopher John Stuart Mill (1806–1873) in *A System of Logic*: "The Law of Causation, the recognition of which is the main pillar of inductive science, is but the familiar truth that invariability of succession is found by observation to obtain between every fact in nature and some other fact which has preceded it (…)."

A related idea is that of determinism, which amounts, in simple terms, to the possibility of predicting future events. In its nineteenth century version, it is embodied in the famous statement by the French savant Pierre-Simon Marquis de Laplace (1749–1827), according to whom[18] "An intelligence knowing, at any given instant of time, all forces acting in nature, as well as the momentary positions of all things of which the universe consists, would be able to comprehend the motions of the largest bodies of the world and those of the smallest atoms in one single formula, provided it were sufficiently powerful to subject all data to analysis; to it, nothing would be uncertain, both future and past would be present before its eyes."

The mechanical determinism exemplified in Laplace's words resulted from the compelling power of Newtonian mechanics in making long-range predictions of many phenomena, particularly in astronomy. The universal scope of these predictions, however, has been denied by the development of deterministic chaos theory in the twentieth century, mainly after the 1960s. In this theory, physicists have demonstrated that for a large variety of physical systems, a small uncertainty in the initial conditions leads, after some time

[17] Albert Einstein, *Out of My Later Years*, quoted by Victor Guillemin [39].

[18] Pierre-Simon Laplace, quoted by Victor Guillemin [42].

evolution, to a broad range of values of the relevant variables (such as position and velocity), making prediction in these cases virtually impossible. Even with phenomena described by simple, well-known physical laws, predictions may be impossible.[19] Although this is different from the fundamental impossibility derived from quantum mechanics, the practical consequences may be analogous.

One of the pioneers of chaos theory was a meteorologist, Edward Norton Lorenz (1917–2008), who found out that the Earth's atmosphere is an example of a system whose behavior is essentially impossible to forecast. The unpredictable influence of one single factor on the evolution of global weather has been dubbed 'the butterfly effect,' after his 1979 paper presented at the American Association for the Advancement of Science, whose provocative title was: "*Predictability: Does the Flap of a Butterfly's Wings in Brazil Set Off a Tornado in Texas*?" [44] He chose the butterfly inspired by another passage of the same Bradbury short story quoted in the footnote given below.

The debate on the philosophical impact of quantum mechanics, in particular on the issues of determinism and causality, still goes on. The uncertainty, non-locality, and probabilistic aspects of the quantum view seem to be inherent to the physical world, not to limitations in the observer's knowledge. In spite of the extraordinary accomplishments of quantum theory, its meaning remains as puzzling as ever. In recent times, some of the counterintuitive aspects of the quantum world, such as the superposition of states (e.g., electrons with spin both up and down), are regarded, rather than as problem, as promising features that open the way to revolutionary new applications. Among these, one finds the great promises of a supercomputer using the inherent parallelism of quantum computation and an unbreakable code using quantum cryptography [45].

[19] This amplification of effects with time can be illustrated with a passage of a short story by the American writer Ray Bradbury, "*A Sound of Thunder*." Speaking of the consequences of a careless time traveler killing a single mouse in the past, and therefore eliminating the following generations of mice, the guide argues: "Well, what about the foxes that'll need those mice to survive? For want of ten mice, a fox dies. For want of ten foxes a lion starves. For want of a lion, all manner of insects, vultures, infinite billions of life forms are thrown into chaos and destruction. Eventually it all boils down to this: fifty-nine million years later, a caveman, one of a dozen on the *entire world*, goes hunting wild boar or saber-toothed tiger for food. But you, friend, have *stepped* on all the tigers in that region. By stepping on *one* single mouse. So the caveman starves. And the caveman, please note, is not just *any* expendable man, no! He is an *entire future nation*. From his loins would have sprung ten sons. From *their* loins one hundred sons, and thus onward to a civilization. Destroy this one man, and you destroy a race, a people, an entire history of life. It is comparable to slaying some of Adam's grandchildren. The stomp of your foot, on one mouse, could start an earthquake, the effects of which could shake our earth and destinies down through Time, to their very foundations." (Ray Bradbury [43]).

Can We Explain Magnetism?

The unveiling of the structure of the atom and the discoveries of quantum mechanics, in particular the idea of the exchange interaction introduced by Werner Heisenberg and Paul Dirac in 1926, explained the mystery of magnetic order in magnetic materials. But what do we really mean when we say that magnetism was 'explained' by these discoveries?

We have seen in earlier chapters how science began and developed from the first efforts of the Greek philosophers, from the pioneering work of the Milesians to the Stoics. This process incorporated different elements of Greek thought, such as the importance attributed by the Pythagoreans to the mathematization of the description of nature in Platonism. In addition, it included the development of the principles of logic by the Stoics and Aristotle and, finally, the value of observation by Aristotle and the seeds of experimentation by Archimedes.

The Greek philosophers put forward several ideas to explain magnetism, these "very hard to understand" effects, in the words of the Chinese *Huai Nan Tzu* treatise. These ideas were usually based on mechanistic elements, like those reasonings that involve the flow of air or the mediation of a vacuum in magnetic attraction. Some descriptions had a vitalistic or animistic character, making the magnet endowed either with life or with a soul. This could not be otherwise, since the Greeks lacked the knowledge of fundamental physical facts that would only be discovered some twenty centuries later. Essentially, these basic facts concerned, on the one hand, the structure of matter and, on the other, the relationship between magnetic and electric phenomena, in both cases, knowledge that would be acquired during the nineteenth and early twentieth centuries. The physics of subatomic phenomena—quantum physics—discovered in the first half of the twentieth century, constituted another aspect of the necessary foundation for the explanation of magnetism.

One feels that human beings have an innate urge to "explain" things; there is a feeling of relief when an explanation is provided for any phenomenon hitherto not understood. A statement that fits an otherwise incomprehensible fact or event into an accepted framework of assumptions, therefore eliminating mystery, is an explanation that produces assurance [46]. "I believe that examination will show that the essence of an explanation consists in reducing a situation to elements with which we are so familiar that we accept them as matter of course so that our curiosity rests," wrote the American physicist and exponent of the operational approach in philosophy of science Percy Williams Bridgman (1882–1961).[20]

What would be the requirements to be fulfilled by a scientific explanation of magnetism? To "explain" this phenomenon, it would be necessary to relate the known empirical facts of magnetism to fundamental physical phenomena or to general laws of nature. For example, the explanation of gravity was given when Newton established the universal Law of Gravitation, i.e., when he identified as a universal property of matter the attraction between objects with a force proportional to their masses.

However, philosophers of science disagree on the possibility of science "explaining" anything at all.[21] For example, in his celebrated work *The Structure of Science* (1961), the Hungarian-born American philosopher Ernest Nagel (1901–1985) writes [49]: "(…) even if the laws and theories of science are true, they are no more than logically contingent truths about the relations of concomitance or the sequential orders of phenomena. Accordingly, the questions which the sciences answer are questions as to *how* (in what manner or under what circumstances) events happen and things are related. The sciences therefore achieve what are at best only comprehensive and accurate systems of *description*, not of explanation."

This is not, however, what scientists, in general, believe and desire to achieve. Scientists are not satisfied until they feel they have discovered what they regard as the "explanations" for the empirically observed facts. A view with which most scientists would identify themselves is that of W. C. Salmon, who sees scientific explanations as statements that show how the given events "fit into the causal structure of the world."[22] Or that of the German-born American philosopher Carl Hempel (1905–1997), according to whom an explanation for an event is given both by "(i) particular effects and (ii) uniformities expressed by general laws, from which the event is to be expected."[23] The latter is very close to the view expressed by the Irish philosopher and bishop George Berkeley (1685–1753) some three hundred years ago[24]: "(...) explication consists only in showing the conformity any particular phenomenon has to the general laws of nature or, which is the same thing, in

[20] P. W. Bridgman, "The Logic of Modern Physics," p. 37, quoted by B. d'Espagnat [47].

[21] It seems wiser not to elaborate much on this theme here, since according to R. Torretti, the discussion of the true meaning of "explanation" gave rise to "a vast and most boring literature". Torretti [48].

[22] W. C. Salmon, in "A Third Dogma of Empiricism," in R. Butts and J. Hintikka, eds., *Basic Problems in Methodology and Linguistic*, 149–166, Dordrecht, D. Reidel, 1977, p. 166, quoted by W. C. Salmon [50].

[23] C. G. Hempel in "Explanation in Science and History," in R. G. Colodny, ed., *Frontiers in Science and Philosophy*, 7–34, University of Pittsburgh Press, Pittsburgh, 1962, p. 10, quoted by W. C. Salmon [50].

[24] G. Berkeley, "A Treatise Concerning the Principles of Human Knowlege [sic]." Dublin, Printed by A. Rhames, for J. Pepyat, 1710, §62, quoted by R. Torretti [51].

discovering the *uniformity* there is in the production of natural effects." These three authors, the first two from the twentieth century and the third from three centuries before agree that science *can* explain the facts of nature. However, they disagree on the importance of causes. The quotations from Berkeley and Hempel are in line with a remark [52] of the French positivist philosopher Auguste Comte (1798–1857), who "proclaimed that explanation by laws, not causes, was the distinctive feature of mature, "positive" science."

How far can we go when we ask "why"? For example, we have seen in the dispute between Leibniz and Newton that they disagreed on the need to search for the origin of gravitation (Chap. 4). Can we ask, in the case of gravitation, why matter has this property of attraction?[25] The answer is that in the modern era, it is not regarded as belonging to the province of physics to find answers to these ultimate "whys." The Greek thinkers asked many "whys" that belonged then and still belong today to the realm of philosophy, or to the branch of philosophy called metaphysics; these were the more general questions like "What is being?" and so on. Some of these are similar to "questions children ask but which adult consciousness first dismisses as unanswerable and then forgets" [54]. Many of the questions the early philosophers asked—those more specific—have been answered by the individual sciences; this is the case with many of the questions related to the physical world.

Moving up to another level of still more general "whys," one can finally ask the question, "Why is there anything, rather than nothing?" This is obviously a very general question; in fact, it is difficult to think of a more all-embracing interrogation than this one, which has been called the super-ultimate question [53]. This issue, maybe the biggest metaphysical problem, has been qualified by many philosophers—perhaps surprisingly, not by all of them—as a meaningless question. Others have described this question as "extraordinary," or a "mystery," or like the German twentieth-century philosopher Martin Heidegger (1889–1976), a question "incommensurable with any other."[26]

"Extraordinary," or a "mystery," this puzzling question remains to this day a very different question from the countless "whys" and "hows" that scientists ask every day about nature and from which scientific progress results. The answers to these queries that relate the problem under scrutiny to fundamental knowledge of the corresponding particular science are, for most practical purposes, valid explanations.

---

[25] On this question, the English mathematician Karl Pearson (1857–1936) wrote in his classic *The Grammar of Science* that we can "describe how a stone falls to the earth, but not why it does" (*The Grammar of Science*, Everyman, 1937, p. 103, quoted by P. Edwards [53]).

[26] M. Heidegger, *An Introduction to Metaphysics*, New Haven, 1959, p. 4, quoted by P. Edwards [55].

References

1. Keith, S. T., & Quédec, P. (1992). Magnetism and magnetic materials. In L. Hoddeson, E. Braun, J. Teichmannn, & S. Weart (Eds.), *Out of the crystal maze, chapters from the history of solid-state physics* (p. 422). Oxford University Press.
2. Sarton, G. (1953). *A history of science* (p. 253). G. Cumberlege.
3. Barnes, J. (1987). *Early Greek philosophy* (p. 248). Penguin.
4. Barnes, J. (1987). *Early Greek philosophy* (p. 247). Penguin.
5. Lucretius Carus, De Rerum Natura. (1971). In E. M. P. Crosland (Ed.), *The Science of Matter* (p. 40). Penguin.
6. Berryman, S. (2022). Ancient Atomism. In E. N. Zalta & U. Nodelman (Eds.), *The Stanford Encyclopedia of Philosophy* (Winter 2022 Edition). https://plato.stanford.edu/archives/win2022/entries/atomism-ancient/
7. Descartes, R. (1971). Discourse on Method, Optics, Geometry and Meteorology.," translated by P. J. Olscamp, Bobbs-Merrill, 1965, pp. 264–5. In M. P. Crosland (Ed.), *The science of matter* (p. 71). Penguin.
8. Descartes, R. (1971). Philosophical writings." translated by E. Anscombe and P. T. Geach, Nelson, 1954, p. 207. In M. P. Crosland (Ed.), *The science of matter* (p. 71). Penguin.
9. Boyle, R. (1971). "The Sceptical Chymist," 1661, modern edn., Dent, 1911, pp. 30–31. In M. P. Crosland (Ed.), *The science of matter* (p. 74). Penguin.
10. Dalton, J. (1971). In M. P. Crosland (Ed.), *The science of matter* (p. 202). Penguin.
11. Scerri, E. R. (1998). *The evolution of the periodic system* (p. 56). Scientific American, September 1998.
12. Thomson, J. J. (1971). In M. P. Crosland (Ed.), *The science of matter* (p. 356). Penguin.
13. Rhodes, R. (1986). *The making of the atomic bomb* (p. 49). Penguin Books.
14. Guillemin, V. (1968). *The story of quantum mechanics* (p. 33). Charles Scribner's Sons.
15. Dyson, F. (1988). *Infinite in all directions* (p. 41). Penguin Books.
16. Dyson, F. (1988). Manchester and Athens. In *Infinite in all directions* (p. 35). Penguin Books.
17. Kragh, H. (1999). Quantum generations, *a history of physics in the twentieth century* (p. 3). Princeton.
18. Muldawer, L. (1969). Resource letter XR-1 on X-rays. *American Journal of Physics, 37*(2), 124.
19. Rechenberg, H. (1995). Quanta and quantum mechanics. In L. M. Brown, A. Pais, & B. Pippard (Eds.), *Twentieth century physics* (Vol. I, p. 150). Institute of Physics Publishing.
20. Heilbron, J. L. (1986). *The dilemmas of an upright man—Max Planck as spokesman for German science* (p. 34). University of California Press.

21. Heilbron, J. L. (1986). *The dilemmas of an upright man—Max Planck as spokesman for German science* (p. 98). University of California Press.
22. Heilbron, J. L. (1986). *The dilemmas of an upright man—Max Planck as spokesman for German science* (p. 24). University of California Press.
23. Whitrow, G. J. (Ed.). (1967). *Einstein, the man and his achievement* (p. 89). Dover.
24. Pyenson, L. (1985). *The young Einstein: The advent of relativity* (p. 8). Adam Hilger Ltd.
25. Pais, A. (1982). *Subtle is the Lord, the science and life of Albert Einstein* (p. 245). Oxford University Press.
26. Rechenberg, H. (1995). Quanta and Quantum Mechanics. In L. M. Brown, A. Pais, & B. Pippard (Eds.), *Twentieth Century Physics* (Vol. 1, p. 146). Institute of Physics Publishing.
27. Stuewer, R. H. (1998). History and physics. *Science & Education, 7*, 13–30.
28. Pais, A. (1995). Introducing atoms and their nuclei. In L. M. Brown, A. Pais, & B. Pippard (Eds.), *Twentieth century physics* (Vol. I, p. 95). Institute of Physics Publishing.
29. Wheaton, B. R. (1992). *The Tiger and the Shark, Empirical Roots of Wave-Particle Dualism* (p. 306). Cambridge University Press.
30. Bloch, F. (1976). *Physics Today, 29*, 23.
31. Dirac, P. A. M. (1963). *The evolution of the physicist's picture of nature* (Vol. 208, pp. 45–53). Scientific American.
32. Helge Kragh, *Dirac—A scientific biography*, Cambridge University Press, 1991, p. 277.
33. Dirac, P. A. M. (1971). The development of quantum theory. In J. Robert *Oppenheimer Memorial Prize Acceptance Speech* (p. 66). Gordon and Breach.
34. David Mermin, N. (1988). Spooky actions at a distance. In *The great ideas today* (pp. 2–53). Encyclopaedia Britannica.
35. Bub, J. (2000). Indeterminacy and entanglement: The challenge of Quantum mechanics. *British Journal of Philosophy of Science, 51*, 597–615.
36. Zeilinger, A. (2000). Quantum teleportation. *Scientific American, 282*, 32.
37. Tegmark, M., & Wheeler, J. A. (2001). *100 years of quantum mysteries* (Vol. 284, p. 54). Scientific American.
38. Bloch, F. (1976). *Physics Today, 29*, 24.
39. Cushing, J. T. (2000). *Philosophical concepts in physics* (p. 317). Cambridge University Press.
40. Guillemin, V. (1968). *The story of quantum mechanics* (p. 263). Charles Scribner's Sons.
41. Mill, J. S. (1936). *A system of logic* (8th ed., p. 213). Longmans, Green and Co.
42. Guillemin, V. (1968). *The story of quantum mechanics* (p. 280). Charles Scribner's Sons.
43. Bradbury, R. (1990). A sound of thunder. In *Classic stories 1* (p. 214). Bantam Books.
44. Gleick, J. (1988). *Chaos, making a new science* (p. 322). Penguin Books.

45. Singh, S. (1999). *The code book: The science of secrecy from ancient Egypt to quantum cryptography*. Anchor Books.
46. Willey, B. (1967). *The seventeenth century background* (pp. 10–11). Penguin.
47. D'Espagnat, B. (1989). *Reality and the physicist* (p. 89). Cambridge University Press.
48. Torretti, R. (1999). *The philosophy of physics* (p. 244). Cambridge University Press.
49. Nagel, E. (1979). *The structure of science* (p. 26). Hackett Publishing Company.
50. Salmon, W. C. (1984). *Scientific explanation and the causal structure of the world* (p. 19). Princeton.
51. Torretti, R. (1999). *The philosophy of physics* (p. 102). Cambridge University Press.
52. Torretti, R. (1999). *The philosophy of physics* (p. 103). Cambridge University Press.
53. Edwards, P. (1967). Why. In P. Edwards (Ed.), *The encyclopedia of philosophy* (p. 297). Macmillan.
54. Willey, B. (1967). *The seventeenth century background* (p. 19). Penguin.
55. Edwards, P. (1967). Why. In P. Edwards (Ed.), *The encyclopedia of philosophy* (Vol. 297, p. 301). Macmillan.

Further Reading

Davies, P. C., & Brown, J. R. (Eds.). (1993). *The ghost in the atom: A discussion of the mysteries of quantum physics*. Cambridge University Press.

Edwards, P. (1967). Why. In P. Edwards (Ed.), *The Encyclopedia of philosophy*. Macmillan.

Guillemin, V. (1968). *The story of quantum mechanics*. Charles Scribner's Sons.

Heilbron, J. L. (1986). *The dilemmas of an upright man—Max Planck as spokesman for German science* (Vol. 55). University of California Press.

Keith, S. T., & Quédec, P. (1992). Magnetism and magnetic materials. In L. Hoddeson, E. Braun, J. Teichmann, & S. Weart (Eds.), *Out of the crystal maze: Chapters from the history of solid-state physics*. Oxford University Press.

Kragh, H. (1999). *Quantum Generations, a history of physics in the twentieth century*. Princeton University Press.

Pais, A. (1995). Introducing atoms and their nuclei. In L. M. Brown, A. Pais, & B. Pippard (Eds.), *Twentieth century physics* (Vol. I). Institute of Physics Publishing.

Stevens, K. W. H. (1995). Magnetism. In L. M. Brown, A. Pais, & B. Pippard (Eds.), *Twentieth century physics* (Vol. II). Institute of Physics Publishing.

Tegmark, M., & Wheeler, J. A. (2001). *100 years of quantum mysteries* (p. 54). Scientific American.

7

Magnets: Large and Small

"One influence of the sunspots upon the Earth is perfectly demonstrated. When the spots are numerous, magnetic disturbances... are most numerous and violent upon the Earth... The nature and mechanism of the connection is as yet unknown, but of the fact there can be no question."
[1]

Summary Electrons and galaxies are physical objects whose sizes differ by a factor of 10^{36}; both produce magnetic fields in their neighborhood. The Sun, other stars, and the terrestrial globe are also magnetically active, and this property is explained by the dynamo model, based on the rotation of a magnetic body. The field produced by the Sun extends for thousands of kilometers and can be felt on the Earth. Some neutron stars have a much higher field than the Sun, the highest found anywhere in the universe, in the range of 10^{12} T (or 10^{16} G); these are the magnetars. The Earth field is felt by the living beings on the planet, who, in some cases, make use of this field for orientation; these include bacteria, birds, and fishes. In the case of bacteria, it was found that they contain microscopic magnetic particles, whose magnetization is oriented by the Earth's field. A phenomenon that has been used for many applications is that of nuclear magnetic resonance (NMR), where an electromagnetic wave induces transitions in the system nucleus-magnetic field. Applications of NMR include nondestructive analysis of chemicals and applications in medicine, such as magnetic resonance imaging (MRI) and functional magnetic resonance imaging (fMRI).

A. P. Guimarães, *A Longstanding Attraction*, https://doi.org/10.1007/978-3-032-02006-2_7

From the end of the sixteenth century, when Gilbert attributed to the magnetism of the Earth the orientation of the compass, it followed the conclusion that objects in widely different scales—the magnetic needle and the planet Earth—showed magnetic "activity." In modern times, it was found out that this scale is much wider since most subatomic particles, as well as galaxies, are surrounded by magnetic fields.

In the microworld, the magnetism of the particles that are constituents of matter, or of those that are only observed in collisions induced in particle accelerators, is either related to the intrinsic angular momentum, or spin, or to their orbital motion, as it is the case, for example, of the electrons moving around the atomic nucleus. The most important particle from the point of view of magnetism is the electron, since, as we have seen, it provides the main contribution to the magnetism of matter. It is also the motion of electrons on a macroscopic scale through wires and coils that creates magnetic fields in electromagnets and solenoids.

Besides the Earth, other astronomical bodies have associated magnetic fields, particularly the Sun and some planets. Our Milky Way and other galaxies also manifest magnetic properties that were revealed through optical techniques, i.e., techniques that extract information on magnetic fields through the analysis of the light emitted from these bodies. In particular, spiral galaxies show a pattern of fields that have been described as "magnetic arms," magnetic field lines that are located between the spiral distribution of stars.

Electrons and galaxies are physical objects that differ in size by a factor of 10^{36} (radius of particle 10^{-15} m, radius of our galaxy 10^{21} m), or a trillion trillion trillion times, that is, a number with 36 zeros. This is a fantastic range of scales in which magnetic phenomena are observed.

Magnetic activity is not only experienced in the inanimate world. On our planet, life has evolved immersed in the local magnetic field, and somehow, many living creatures sense and make use of this field for orientation.

The Great Magnet

"Magnus magnes siue terrestris globus," "The great magnet or the terrestrial globe,"– stated Gilbert in 1600, in the *De Magnete*.[1] This discovery is remarkable in many ways and represents an important step in the knowledge about

[1] It has been argued that the sentence often attributed to Gilbert ("Magnus magnes ipse est globus terrestris") is not to be found in his work but was a version due to Von Humboldt [2].

the Earth; in fact, this magnetic property amounts to the second overall attribute to be associated with our planet, the first one being its roundness [3].

Although it is, in fact, a great magnet, the Earth is not like a huge piece of magnetite; its magnetism is more like that of an electromagnet, arising from macroscopic electric currents, as will be explained below.

The magnet Earth creates in its neighborhood, and especially on its surface, a magnetic field that has many observable effects; the power to align the magnetic needle was the first of these effects to be discovered (Fig. 7.1). The characteristics of this field have been studied by Gilbert and by uncountable researchers in the ensuing four hundred years: its intensity, direction relative to the surface of the planet, variation with time, and so on. The magnetic properties of the Earth could only begin to be explained after the relationship of electricity and magnetism had been unveiled in the first half of the nineteenth century.

As we have already discussed in Chap. 3, the magnetic field of the Earth, in general, deviates at each point of the surface from the direction of the meridian—this deviation is called declination. At the end of the fifteenth century, western European sailors knew that the compass pointed east of true north. In September 1492, the sailors under the command of Christopher Columbus (c.1450–1506) on his first voyage were "afraid" and "dismayed" when, west of the Azores, the magnetic needle shifted to a westward declination. This meant that the declination changed from point to point on the surface of the Earth, and Columbus' entry in his journal on September 13, 1492, is the first report of this westward declination in the western Atlantic [4].

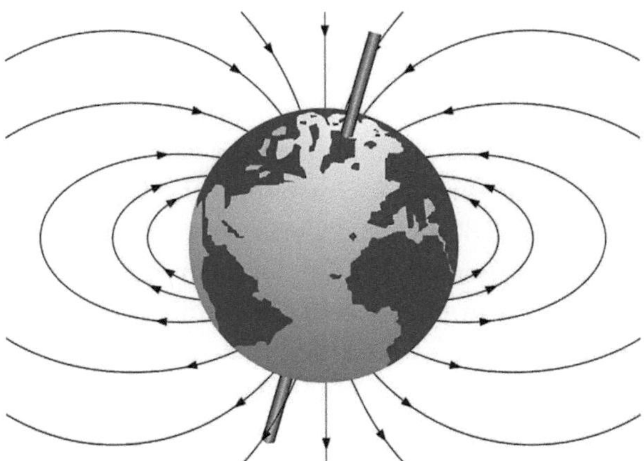

Fig. 7.1 Earth magnetic field lines

The first worldwide study of declination was performed by the Portuguese naval commander and scholar João de Castro (1500–1548), who sailed to the East Indies and to the Red Sea in 1538, making some fifty observations [5]. He derived the direction of true north from the position of the sun and compared it with the direction of the magnetic compass, thereby deriving the declination at each point. Castro was later appointed viceroy of Portuguese India, where he died in Goa in 1548.

In the following century, in 1698 and 1700, the English astronomer and mathematician Edmond Halley (1656–1742) compiled the first chart of values of declination, with data collected in an expedition that constituted the first sea voyages organized for purely scientific purposes. His *Atlantic Chart* was published in 1701, and it represented the magnetic declination by means of contour lines. This is the first known example of the representation of a variable through contour lines, a type of graph subsequently called a "Halleyan" curve [6].

The deviation of the magnetic needle from the horizontal plane—the inclination, or dip—was first studied systematically by Gilbert, who found that the compass pointed horizontally at the Equator and turned gradually away from the horizontal plane as one moved towards the Poles. Gilbert initially thought that the angle between the compass and the plane gave the latitude directly, but this is only a very rough measure.

A magnetized needle that is displaced from the direction of the magnetic field and then released will tend to realign itself with the field, oscillating until it reaches parallelism with it. The needle will oscillate more rapidly the stronger the intensity of the field. This fact has been used to measure the intensity of the Earth's field by deriving it from the period of oscillation. The first survey of magnetic field intensity using this technique was made by the French Robert de Paul de Lamanon (1752–1787) in the expedition led by Jean François Galaup, count de La Pérouse (1741–c. 1788) [3], a French navigator. Unfortunately, the data were lost with the sinking of his ship, *La Boussole*. In 1791–1794, Admiral De Rossel determined the field intensity as a function of latitude, verifying the systematic increase from the Equator to the Poles; his results were published in 1808. This dependence of the magnitude of the field on latitude was also observed by the German naturalist Alexander von Humboldt (1769–1859) in his scientific mission of 1799–1803. He valued these findings so much that he described this accomplishment as "the most important result" of his American voyage [7].

The geomagnetic field tends to be more intense near the poles, with a maximum of about 6/100,000 tesla (or 0.6 gauss) near the poles and minima near the equator; the lowest value, of about 2.4/100,000 tesla, is found on the South American continent, near Rio de Janeiro, Brazil [8].

The magnetic field of the Earth also exhibits a variation with time, both in intensity and in declination. In 1634, the English astronomer Henry Gellibrand (1597–1636), a professor at Gresham College in London, discovered the variation of declination with time. This effect has since been followed for more than three hundred years. The English watchmaker George Graham (1674–1751), pointing a microscope at the tip of a magnetic compass, detected some very rapid oscillations that occur in a matter of seconds due to minute changes in declination.

Variations of the intensity of the magnetic field as a function of time are observed on different time scales. There are changes that occur on a very short timescale, in the range of thousandths of seconds, and there is a slow variation, on the scale of millions of years. On the long- time scale, it is known that the intensity of the field has been decreasing during the last thousands of years at a rate of about 5% per century and is therefore expected to reach a value of zero around the year 4000 if this tendency persists.

The direction of magnetization of the planet Earth has also changed many times in the past; in the last six million years, it has reversed approximately once every 300,000 years.

The knowledge of the long-term variations of the Earth's magnetic field is mainly based on the study of the magnetic properties of rocks. The principle is simple: igneous rocks, formed, for example, from lava flow, if solidified under a magnetic field, will show some residual magnetization, and this can be related both in intensity and direction to the extant magnetic field. Sedimentary rocks, formed under the same conditions of a magnetic field, also retain a magnetization. This magnetization is called, in general, remanent magnetization (thermoremanent, in the first case). Another mechanism of magnetization arises from the enormous electric currents that flow when, during a thunderstorm, lightning reaches the ground.

The Irish physicist Joseph Larmor (1857–1942) proposed in the twentieth century (1919) a theory to explain the origin of the magnetism of the Sun; this is called the dynamo model. The theory was later applied to the Earth and has been refined by several researchers, remaining the basis of the physical explanation of the Earth's magnetism. It is attributed to the rotating molten iron core of the Earth, the role of producing, like a dynamo, an electric current that is responsible for the magnetic field. Long-period changes in the complex motion of the magma account for the variations in the characteristics of the geomagnetic field.

The magnetic field in the neighborhood of the Earth affects the motion of charged particles emitted by the Sun (mostly protons and electrons), which constitute the so-called "solar wind." The influence of the field effectively traps these particles, creating charged clouds at a distance of thousands of kilometers from the Earth, called Van Allen belts. Explosions in the sun called flares emit large surges of particles that reach the Earth after traveling for about 2 days. When they attain the upper atmosphere, they produce a phenomenon designated magnetic storm, in which the magnetic field measured on the Earth's surface is modified and which may lead to the disruption of shortwave communications.

The spectacular display of the aurora borealis, or northern lights, was first attributed to the Earth's magnetic field by Edmond Halley in the seventeenth century [9]; it is known today that the lights arise from the effects of the solar wind particles captured by this field as they hit the outer layers of the atmosphere. The same phenomenon occurs in the southernmost latitudes and receives the name of aurora australis, or southern lights.

The Moon is also a magnet, but much weaker than the Earth; the measure of its strength, the so-called magnetic moment, is 100 million times smaller than that of our planet. The small resulting magnetic field measured at the Moon's surface is due to remanent rock magnetism [10]. There is no active dynamo mechanism in our satellite. Jupiter and Saturn, on the other hand, have strong magnetic properties associated with the dynamo effect; Jupiter has a magnetic moment ten thousand times bigger than the Earth's.

Uranus and Neptune have uncommon magnetic properties. Both planets present magnetic moments that are tilted by large angles (of about 50 degrees) in relation to their axes of rotation. The source of magnetism of these planets does not seem to be located near the planet's center.

Many other astronomical bodies are magnetically active; the most important example for us is that of the Sun, which produces a field that extends in space over huge distances, reaching the orbit of the Earth, with a value at this distance 10,000 times smaller than the maximum field due to our own planet.

Magnetic fields play a very important part in solar phenomena, the most striking of which is the cycle of sunspots, darker regions that appear this way since they have temperatures more than one thousand degrees lower than the visible surface of the Sun. The invention of the telescope led to the discovery of sunspots, first reported in 1609 by Galileo and other observers. The German pharmacist and amateur astronomer Samuel Heinrich Schwabe (1789–1875) published in 1843 evidence for the existence of a sunspot cycle [11]: every 11 years, the number of sunspots goes through a maximum and decreases again. It was later established that every two 11-year cycles, the north and south magnetic poles of the Sun change polarity.

The overall magnetism of the Sun arises from essentially the same dynamo mechanism that is active on Earth, with the extra complexity introduced by the fact that the ionized (electrically charged) gases in the Sun are conductors of electricity. Sunspots exhibit magnetic fields thousands of times stronger than those observed on the Earth's surface.

Magnetic fields in stars are also due to the dynamo mechanism. These fields can be detected through their effects on the emitted light and radio waves; this also allows an estimate of the field intensities. Some stars are known to be relatively small objects, with a radius of about 10 km; these are neutron stars, stars of ultra-dense matter, formed practically only of neutrons. These strange bodies have very high magnetic fields, of some 10^{10} T (or 10^{14} G), or one hundred trillion times (10^{14}) higher than Earth fields. Some neutron stars have an even higher field, the highest field found anywhere in the universe, with a magnitude one thousand trillion times (10^{15}) the Earth's field. This class of stars is called magnetars [12]. These stars were observed for the first time in 1979 when one neutron star produced an ultra-powerful pulse of X-rays, so powerful that in the first two-tenths of a second it emitted more energy than the Sun radiates in one thousand years. The magnetic fields on their surface are typically of the order of 10^{15} G (or 10^{11} T), and internal fields of the order of 10^{17} G (or 10^{13} T) (Table 7.1).

The presence of magnetic fields has also been detected in distant galaxies and even in intergalactic space. The fields in the spiral galaxies tend to be stronger between the spiral arms and parallel to them, with a low intensity, some 100,000 times smaller than the field at the Earth's surface. Their origin is not completely understood; they are probably due to an amplification of original fields that had arisen in the early moments after the Big Bang. This amplification could result from a dynamo mechanism, as the one that acts in the Earth's core; some authors believe that the presence of these original magnetic fields was important for the shaping of galaxies [13].

Table 7.1 Some values of magnetic fields, in tesla (1 T = 10^4 gauss)

Magnetars	10^{12} T
Neutron stars	10^{8} T
White dwarf stars	10^{4} T
Sunspots	1 T
Earth surface	10^{-4} T
Milky way	10^{-8} T
Heart	10^{-8} T
Brain	10^{-12} T

Electrons and nuclei of high energy—the cosmic rays—impinge regularly on the Earth's high atmosphere. These particles, with velocities approaching that of light, collide with nuclei in the atmosphere and produce showers of secondary particles that can be detected at sea level. The acceleration of ultra-high-energy cosmic rays, one of nature's deepest secrets, may be explained by the action of magnetic fields outside our own galaxy [13].

From immense galaxies down to a small lodestone, magnetic properties vary enormously in scale; however, these always arise either from the intrinsic moments or from the motion of electrically charged particles.

Living Magnets

The extraordinary ability some birds have to find their way, either migrating for very long stretches or returning to their shelter when released a considerable distance away, has been known for a long time. Before our era, pigeons were used by Julius Caesar (c. 100 BC–44 BC) to send to Rome the news of the conquest of Gaul. Genghis Khan (c. 1160–1227), in the thirteenth century, also employed pigeons as messengers as his army progressed into Eastern Europe. The earliest well-documented case is the use of pigeons for a period of 6 months during the siege of Paris, in the 1870–1871 Franco-Prussian War [14]. The birds were taken out of Paris by balloons, and the messages were photographically reduced; one bird once carried more than 40,000 messages. This use of pigeons continued into the twentieth century, when they were still employed in the First and Second World Wars.

In the nineteenth century, investigators began the scientific study of migrating birds and homing birds, trying to determine whether the capacity of these birds to navigate over unknown terrain was related to the use of some internal magnetic compass [14].

Living organisms, since their appearance on our planet, some four billion years ago, have existed in environmental conditions that include the presence of the geomagnetic field. Although this field has changed both its intensity and direction on such an extended time scale, it may have played a role in the preservation of life on Earth. This is so because the charged particles of the solar wind tend to be trapped in the Earth's magnetic field in space, therefore reducing their flow to the surface of the planet. Such reduction probably contributed to the survival of the first living organisms.

Beginning in the nineteenth century, investigators tried to determine the effect of magnetic fields on living creatures. One of the earliest of such studies was performed in the laboratory of the American inventor Thomas Edison

(1847–1931) in 1892: the largest electromagnet existing at that time was used to apply a field of 0.15 T to a boy and a dog, with no apparent effect.[2] Another line of inquiry has been the attempt to find out the influence that the Earth's magnetic field might have on the behavior of living beings. Is the ability of birds that travel for thousands of miles somehow related to the geomagnetic field? Do they have some form of internal magnetic compass?

Through experiments made with applied magnetic fields, data on magnetic orientation, obtained especially since the 1960s, have shown that many living creatures, both invertebrate and vertebrate, have this type of ability; among the latter are fish, amphibians, reptiles, birds, and mammals. Investigations with human subjects have found a marginal response to the influence of environmental magnetic fields, and the results are controversial [16].

Many studies have demonstrated that pigeons can find their way using visual cues, using the position of the sun, as well as information from the geomagnetic field. Under overcast skies, the importance of magnetic data becomes fundamental for their guidance. Experiments have shown that the navigating ability of the birds can be hindered by increases in magnetic activity in the atmosphere, as those occurring during magnetic solar storms. In addition, it has been shown that pigeons with magnets attached to their bodies may be disturbed in their flight back to the coop.

To accomplish their navigation exploits, pigeons need to define a direction, i.e., they require what one might call a compass sense and also determine their position (map sense). The findings point to the use of magnetic compass information by the birds. However, the relevance of magnetic factors for their map building, which makes use of the variation in inclination (or dip) as a function of latitude and variations in the intensity of the magnetic field, is not so well established [17].

In vertebrates, two mechanisms of orientation seem to act: in birds and sea turtles, the inclination of the magnetic field is used to define "poleward" (i.e., the direction towards the pole, indifferently to which pole, north or south) as the direction along which the angle between the vertical and the field direction is smaller. On the other hand, in salmon and mole rats, the internal compass distinguishes north from south, and their navigation uses this knowledge [18].

Honeybees make use of magnetic information in their foraging excursions and also employ it to align their combs. The behavior of other animals, such as eels and newts, also shows evidence of magnetic orientation.

---

[2] F. Peterson and A. E. Kennely, *Some Physiological Experiments with Magnets at the Edison Laboratory*, NY Medical Journal 56 (1892) 729, quoted by F. Peterson and A. E. Kennely [14].

The simplest living beings that show magnetic orientation are bacteria; this behavior of the bacteria was observed for the first time by Salvatore Bellini (University of Pavia, Italy) and reported in 1963 [19].

This property was rediscovered in 1975, by accident, by R. P. Blakemore, then a graduate student of microbiology at the University of Massachusetts. For these bacteria, which were named *Magnetospirillum Magnetotacticum*, the living organisms have a passive role, in the sense that the geomagnetic field orients them so that the motion of the bacterial flagella, a hair-like structure that functions like a propeller, conducts them along the lines of the field. These are called "magnetotactic" bacteria, from the root tactic, directed or oriented by some agent. Since in most regions of the planet the field lines point either up or down, this alignment may help bacteria to avoid the oxygen-rich medium near the surface of the ponds and search for nutrition near the bottom. In confirmation of this hypothesis, it is observed that magnetic bacteria in the southern and northern hemispheres have opposite polarities.

The discovery of magnetite in these bacteria, where it has the form of chains of grains of 1/10,000 of a millimeter (or 1/250,000 of an inch) in diameter, stimulated a search for magnetic substances, or "compasses," in the tissues of different animals. Many instances of magnetic substances have been found since then in living creatures, such as honeybees, pigeons, turtles, and salmon. In practically every case, the magnetic substance is formed of very small grains of magnetite synthesized in the organisms through a process known as biomineralization. The mechanism through which the interaction of the magnetic field with the grains of magnetite reaches the central nervous system of the organisms, affecting their behavior, is still not understood.

Magnetic Resonance: Dancing Spins

We have seen that in the same way that an electron carries magnetism, atomic nuclei may also have a "magnetic moment," albeit a much smaller one. Since this nuclear magnetism is much weaker than the normal magnetism of matter, which arises from the electrons, its effects are not easily perceived. The role of nuclei in the magnetism of matter is so small that only refined experimental techniques developed in the mid-twentieth century allowed the detection and analysis of this contribution. The phenomenon that most easily demonstrates the presence of nuclear magnetism is nuclear magnetic resonance (NMR), which constitutes the physical basis of the magnetic resonance imaging technique (MRI).

Atomic nuclei in the presence of an applied magnetic field circle around the field direction like a top, with a frequency of rotation, or precession, that is characteristic of the type of nucleus and is also proportional to the intensity of the magnetic field. If a radio wave of the same frequency is made to impinge on an ensemble of precessing nuclei, it will be absorbed. This phenomenon is resonant, which means that the absorption occurs only if the frequency of the wave is the same as the turning frequency of the nuclei. Therefore, by varying the frequency of the wave and recording the variation of its absorption rate, an experimenter can determine with great precision the frequency of precession.

In the summer of 1940, a British mission led by Sir Henry Tizard (1885–1959) of the British Government Scientific Advisory Committee visited the United States to discuss ways of pooling the scientific resources of the two countries against the German forces in the early days of World War II. They took with them a new device for the production of radio waves that had been developed in the previous year at the University of Birmingham, the "magnetron." In the magnetron, a permanent magnet made the electrons move in circular orbits, in which they emitted electromagnetic waves (microwaves) of much higher power than the previously employed radio sources. The magnetron was vital for the development of radar, a technique that helped the Allies win the war. For this reason, the magnetron was hailed in America as "the most valuable cargo ever brought to our shores."[3]

One of the practical peacetime outcomes of the invention of the magnetron and the technical developments associated with the technology of radar was the design of the microwave oven; this home appliance is the most common use of the magnetron. Another important consequence was the discovery of nuclear magnetic resonance.[4]

The first attempts to observe nuclear magnetic resonance were made [22] by the Dutch physicist Cornelis Jacobus Gorter (1907–1980) in Leiden in 1932. He tried then to detect the resonance through a rise in temperature of the samples on the absorption of the radio waves, but the sensitivity of the experiment was not sufficient to allow a positive result. A second experiment, carried out in 1942 with Lambertus Johannes Folkert Broer (1916–1991) using a different technique, again failed, paradoxically due to an experimental difficulty associated with the fact that the samples used were too pure.

[3] James Phinney Baxter, *Scientists Against Time*, quoted by James Phinney Baxter [19].

[4] The phenomenon of magnetic resonance was first suggested by the Russian physicist J. Dorfmann in 1923. J. Dorfmann [20].

In 1938, the American physicist Isidor Isaac Rabi (1898–1988) of Columbia University published with collaborators an account of the nuclear magnetic resonance effect using a beam of molecules of lithium chloride moving in a vacuum. The first investigator ever to report the experimental observation of magnetic resonance in a condensed phase, in this case not the resonance of nuclei, but of electrons in a chromium salt, was Yevgeny K. Zavoisky (1907–1976), of the Kazan State University, in Russia, in 1944 [23].

The positive observation of nuclear magnetic resonance in solids or liquids was only reported in 1946 by two American groups working independently [24]: Edward Mills Purcell (1912–1997), Henry C. Torrey (1911–1998), and Robert V. Pound (1919–2010), at MIT, and Felix Bloch (1905–1983), William W. Hansen (1909–1949), and Martin Packard (1921–2020) at Stanford University.

Nuclear magnetic resonance (NMR) has been applied for several decades as a powerful analytic tool in chemical studies; this use arises essentially from the fact that the nucleus of the atoms resonates or precesses at slightly different frequencies, depending on its chemical environment. Although nuclei of several elements have magnetic moments and, therefore, can be used in nuclear resonance experiments, most of the NMR chemical studies are made with hydrogen nuclei and a smaller proportion with the carbon isotope of mass 13. The NMR spectra of molecules give information on the number and location of the hydrogen atoms. Modern so-called solid-state NMR has been applied to studies of the dynamics, structure, and morphology of new materials, such as polymers, proteins, glasses, ceramics, and liquid crystals.

In magnetically ordered solids, e.g., in magnets, the NMR technique has been applied in structural studies, exploiting the influence of impurities, structural defects, and changes in magnetic order on magnetic resonance. The NMR spectrum allows the probing of the immediate neighborhood of the resonant nucleus. In these ordered solids, the magnetic moments of the electrons generate large magnetic fields at the nucleus, and thus, nuclear magnetic resonance is observed without the need to apply external magnetic fields [25].

Another promising field of investigation using NMR is quantum information and computation [26]. The possibility offered by the nuclear resonance technique of changing the orientation of nuclear magnetic moments in a magnetic field has suggested that NMR can be used in computing. An ensemble of nuclei with "spin up" or "spin down" can represent a binary number, and since NMR pulses can change their orientation, therefore changing the stored bits, they may perform mathematical operations on these data.

A quantum computer employing the NMR of ^{19}F nuclei demonstrated in 2001 [27] the factoring of a number (15, in this case).

Nuclear magnetic resonance has important applications in medicine, e.g., in the magnetic resonance imaging (MRI) technique (see Chap. 8).

This chapter opened with a discussion of the remarkable assertion that the Earth and many other celestial bodies are huge magnets. As proof of the extraordinary range of magnetic phenomena observed when the facts of nature are studied on widely different scales, the chapter closes with a mention of quantum computers, machines that, someday in the future, may operate based on the magnetism of the atomic nuclei.

References

1. Young, C. A. (1888). *A textbook of general astronomy*. Ginn and Co.
2. van der Sluijs, M. A. (2014). *Eos, 95*(16), 137.
3. Merrill, R. T., McElhinny, M. W., & McFadden, P. L. (1998). *The magnetic field of the earth* (p. 7). Academic Press.
4. Bedini, S. A. (1992). Compass. In *The Christopher Columbus Encyclopedia* (Vol. 1, p. 207). Simon & Schuster.
5. de Albuquerque, L. (1994). Contribuição das Navegações do séc. XVI para o Conhecimento do Magnetismo Terrestre. In *Estudos de História da Ciência Náutica, Ministério do Planejamento e da Administração do Território, Lisboa* (pp. 249–267).
6. David, R. (1989). Barraclough, geomagnetism: Historical introduction. In D. E. James (Ed.), *The encyclopedia of solid earth geophysics* (pp. 584–592). Van Nostrand Reinhold.
7. von Humboldt, A. (1951). *Quoted in Sydney Chapman and Julius Bartels* (p. 913). Oxford University Press, Oxford.
8. Kirschvink, J. L., Jones, B. S., & MacFadden, B. J. (1985). *Magnetite biomineralization and Magnetoreception in organisms* (p. 49). Plenum Press.
9. Wolf, A. (1962). *History of science, technology, and philosophy in the 16th and 17th century* (Vol. 1, 2nd ed., p. 302). George Allen & Unwin.
10. Merrill, R. T., McElhinny, M. W., & McFadden, P. L. (1998). *The magnetic field of the earth* (p. 394). Academic Press.
11. Stern, D. P. (2002). A Millenium of Geomagnetism. *Reviews of Geophysics, 40*(3), 1–30, p. 13.
12. Beskin, V. S., Balogh, A., Falanga, M., & Treumann, R. A. (2016). Magnetic Fields at Largest Universal Strengths: Overview. In V. S. Beskin, A. Balogh, A. M. Falanga, M. Lyutikov, S. Mereghetti, T. Piran, & R. A. Treumann (Eds.), *The strongest magnetic Fields in the universe*. Springer.
13. Widrow, L. M. (2002). Origin of galactic and extragalactic magnetic fields. *Reviews of Modern Physics, 74*, 775.

14. Woods, D. L. (1965). *A history of tactical communication techniques* (p. 63). Orlando.
15. Schenk, J. F. (2000). Safety of Strong, Static Magnetic Fields. *Journal of Magnetic Resonance Imaging, 12*, 2, 5.
16. Wiltschko, R., & Wiltschko, W. (1995). *Magnetic orientation in animals* (p. 71). Springer-Verlag.
17. Wiltschko, R., & Wiltschko, W. (1995). *Magnetic orientation in animals* (p. 170). Springer-Verlag.
18. Lohmann, K. J., & Johnsen, S. (2000). The neurobiology of magnetoreception in vertebrate animals. *Trends in Neuroscience, 23*(4), 1542–1548.
19. Bellini, S. (1963). *Su di un particolare comportamento di batteri d'acqua dolce.* Istituto di Microbiologia dell'Università di Pavia.
20. Ronald, W. (1965). *Clark* (p. 268). MIT Press.
21. Dorfmann, J. (1923). Einige Bemerkungen zur Kenntnis des Mechanismus magnetischer Erscheinungen. *Z Phys, 17*, 98–111.
22. Gorter, C. J. (1967). *Bad luck in attempts to make scientific discoveries* (pp. 76–81). Physics Today.
23. Vonsovskii, S. V. (1974). *Magnetism* (Vol. 1, p. 10). John Wiley.
24. Pound, R. V. (1999). From radar to nuclear magnetic resonance. *Reviews of Modern Physics, 71*, S54.
25. Guimarães, A. P. (1998). *Magnetism and magnetic resonance in solids.* John Wiley.
26. Jones, J. A. (2001). NMR quantum computation. *Progress in Nuclear Magnetic Resonance Spectroscopy, 38*, 325–360.
27. Vandersypen, L. M. K., Steffen, M., Breyta, G., Yannoni, C. S., Sherwood, M. H., & Chuang, I. L. (2001). Experimental realization of Shor's quantum factoring algorithm using nuclear magnetic resonance. *Nature, 414*, 884.

Further Reading

Beskin, V. S., Balogh, A., Falanga, M., & Treumann, R. A. (2016). Magnetic Fields at Largest Universal Strengths: Overview. In V. S. Beskin, A. A. Balogh, M. Falanga, M. Lyutikov, S. Mereghetti, T. Piran, & R. A. Treumann (Eds.), *The Strongest Magnetic Fields in the Universe.* Springer.

Guimarães, A. P. (1998). *Magnetism and magnetic resonance in solids.* John Wiley.

Marmol, M., Gachon, E., & Faivre, D. (2004). Colloquium: Magnetotactic bacteria: From flagellar motor to collective effects. *Reviews of Modern Physics, 96*, 2024.

Wiltschko, R., & Wiltschko, W. (2019). Magnetoreception in birds. *Journal of the Royal Society Interface, 16*, 20190295.

Wiltschko, R., & Wiltschko, W. (1995). *Magnetic orientation in animals.* Springer-Verlag.

8

Magnetism in Medicine: Magnets that Cure

Such as one—perhaps the oldest extant—is the papyrus Ebers dating from about 1500 BC. This 110 page "book" contains such well known drugs as hartshorn and castor oil, with an indication of the disease or complaints for whose cure they were recommended.
[1].

Summary In Egypt, medicine was organized almost 3000 years before our era. The Greeks were influenced by the Egyptians, e.g., the Greek Hippocrates (c. 460–c. 370 BC) studied in the library in Egypt. Medicine was developed in China and India more or less at the same time that it grew in Greece. These two civilizations kept many early practices to this day. Medical instrumentation was created in antiquity and refined over the centuries, allowing the measurement of blood pressure and body temperature and registering the electrical activity of the heart and brain. Records of the use of magnets in medicine are very old: e.g., in India, a magnet helped to extract an arrow tip in 600 BC. Today, several techniques use the magnetic moments of nuclei in magnetic resonance imaging (MRI) and functional MRI (fMRI), employed to obtain images of the organs. Using the magnetic properties of the hemoglobin or magnetic fields generated by active regions of the brain, neural activity can be mapped. Other applications of magnets include guiding instruments such as catheters and using microscopic magnetic particles in diagnoses and therapy. Finally, in the eighteenth century, Friedrich Anton Mesmer (1734–1815) created a treatment using magnets that was denounced as non-scientific.

A. P. Guimarães, *A Longstanding Attraction*, https://doi.org/10.1007/978-3-032-02006-2_8

Early Medicine

The examination of fossils, mainly from the Neolithic period (from about 10,000 years to 2,000 years before our era) or from the end of the Paleolithic [2], shows craniums exhibiting healed trepanations, indicating that these surgical interventions were made on live individuals that survived them [3]. Fossils of the pre-Columbian cultures also show these marks. Apparently, this procedure was supposed to liberate demons responsible for some nervous or mental condition.

In Egypt, medicine developed about 2800 BC, reaching a significant reputation in the ancient East, although it never abandoned magic remedies and practices. The techniques adopted by the Egyptians reached the West through the Greeks. Theophrastos (c. 371—c. 287 BC), Pedanius Dioscorides (c. 40–90 AD), and Galen of Pergamon (129–216 AD) always refer to Egyptian prescriptions learned, according to Galen, consulting the library at the temple of Imhotep, in Memphis, Egypt, where Hippocrates (c. 460–c. 370 BC), known as the 'father of Medicine,' had studied seven centuries before. According to the historian and geographer Herodotus (c. 484—c. 425 BC), each (Egyptian) doctor specialized in treating one malady.[1]

In Homer's *Odyssey* (eighth century BC), the prestige of Egyptian medicine is apparent: Helen's grief for the absence of Ulysses is attenuated by adding to her wine a drug from Egypt, "where each doctor is the wisest of men" and "of the family of Paeon,"[2] Paeon being the doctor of the gods. Later, the Greeks started to attribute the invention of medicine to Asclepius, son of Apollo, who was eventually worshipped as a god [5].

Hippocrates (c. 460–c. 370 BC), was born on the island of Cos in Greece. He made a very significant contribution, changing medicine's practical and conceptual framework. One may consider that he [6] "systematized the empirical knowledge which had accumulated in Egypt and in the schools of Cnidus and Cos, and founded inductive and positive medicine. He did for medicine what Socrates did for philosophy."

Galen of Pergamon (129–216 AD), another giant in the history of medicine, left some twenty thousand pages of his writings. He studied philosophy and mathematics before embracing medicine. He traveled through different centers of the Hellenistic world and finally settled in Rome. The results of his studies of human anatomy were relevant until the Renaissance [7].

---

[1] In this Section, we have drawn extensively from *A History of Medicine, From Prehistory to the Year 2020* Sutcliffe, J. & Nancy Duin, N. [4].

[2] Homer, *Odyssey* (4227–232).

The first steps in establishing scientific elements in medicine occurred more or less in the same period in Greece, China, and India. In Greece, the later evolution of the methods and approaches became part of the body of scientific medicine; in China and India, many elements of the early concepts and practices have survived into our days.

One of the earliest Chinese medicine texts is the *Nei Ching* (*Yellow Emperor's Classic of Internal Medicine*), probably compiled between 479 and 300 BC; it deals mainly with acupuncture.

It is impossible to date the origins of the Indian system of medicine, now known as Ayurveda (literally, "knowledge, or science, of life"); few written records survive. However, the four *Vedas,* sacred Sanskrit poems of the second millennium BC, have been preserved. They were, in the beginning, orally transmitted. The standard medical compendium with contributions of the physician Charaka (probably from the fourth century BC) was the *Charaka Samhita,* which appeared in the first century AD and described 300 bones and 500 muscles. Its nosology, or list of diseases, was equally detailed [8]. Another related text was the *Susruta* (or *Sushruta*) *Samhita* (created sometime between the twentieth century BC and the sixth century AD), which became a bible for surgery, listing 121 types of surgical instruments [9].

Greek medicine was very influential in Rome for many centuries. After Rome had fallen to the Germanic tribes in the fifth century AD, the learning center moved to Constantinople, present-day Istanbul.

During the Middle Ages, many institutions to shelter paupers, including sick people, were created in different places. However, institutions focused on providing medical care—hospitals—seem to have appeared for the first time in Constantinople: the Sampson Hospital (named after a saint of the fourth century) is one of the first to be created, in the seventh century.[3] Monasteries during this period were important centers in the preservation of knowledge, and many works, including those on medicine, were translated. Medical education was expanded and improved. In the twelfth century, the first European medical school was established in Salerno, in the southwest of the Italian peninsula.

In this period, medicine developed in the Arab world with very important contributions from the Persian Avicenna (also known as Ibn Sina) (980–1037), author of the *Canon Medicina*, a treatise (of about a million words) [10] used in Western Europe until the seventeenth century.

In the fourteenth century, the plague, the Black Death, spread to Europe, killing about one-third of the total population. Only in the nineteenth

[3] Lindberg, D. C. (2007), Op. cit., p. 349.

century was the bacillus, found in black rats, identified as the transmitter of the epidemic.

During the Renaissance, the period between the centuries fourteenth and seventeenth, new contributions were made to the advancement of medicine. These include anatomical drawings that made known the shapes of the organs and the first description of the blood circulation in the lungs. One should mention the drawings of Leonardo da Vinci (1452–1519) and of the Professor of Anatomy, Andreas Vesalius (1514–1564).

William Gilbert (1544–1603) mentions in *De Magnete* authors who have prescribed the lodestone, among them Dioscorides, Galen, and Garcias ab Horto. Ever since Gilbert wrote this, many physicians have practiced with magnets; one of the best-known cases was that of the Irish doctor, Valentine Greatrakes (c. 1628-c.1700) [11] in the seventeenth century, reported to have effected many "cures," attracting large numbers of patients [12]. Athanasius Kircher (c. 1602–1680), in *Magnes, sives De Arte Magnetica* ("*The Magnet, or About the Magnetic Art*") (1641), discusses the healing power of the magnet and the importance of magnetism in many natural phenomena; he concludes that God is all nature's magnet (*totius naturae magnes*) [13].

The movement known as the Enlightenment, in the seventeenth and eighteenth centuries, represented a tendency to reject traditional ideas, with an increased valorization of rationalism. In this period, the British doctor William Harvey (1578–1657) discovered that blood flows as—in his words—'the movement of the blood occurs constantly in a circular manner and is the result of the beating of the heart' [14].

Significant advances were also made in medical instrumentation; for example, Anton von Leeuwenhoek (1632–1723), from Delft, Dutch Republic, explored the medical use of the early microscope, visualizing protozoa and bacteria. Another milestone was the invention of the vaccine for smallpox by the English physician Edward Jenner (1749–1823), who published his discovery in 1798; smallpox was considered eradicated by 1980.

In 1758, the Rev. Edward Stone (1702–1768), in Oxfordshire, England, discovered that chewing a twig of the white willow tree (*Salix alba*) relieved his fever. In 1835, the German chemist Karl Jacob Löwig (1803–1890) obtained a substance later known as salicylic acid, which received the commercial name of *aspirin*, a product that also had the property of being a painkiller.

Opium, derived from the white Indian poppy (*Papaver somniferum*), has been used since antiquity, for example, in Sumer, Egypt, and Persia. In the sixteenth century, it was routinely employed in Western European medicine. In the Middle Ages, sponges impregnated with opium were used as anesthesia.

Childbirth can be a moment of great joy, but it has happened frequently in the past to be the opposite, with the death of the mother or the baby after the birth. Ignaz Semmelweis (1818–1865), a young assistant of Hungarian origin at the Vienna lying-in hospital, found a correlation between the high incidence of deaths and the lack of hygiene in one of the hospital wards. After precautions were taken, the loss of lives was significantly reduced.

Louis Pasteur (1822–1895), a French scientist, demonstrated that microbes did not appear from anything, i.e., there was no spontaneous creation, and germs present in the atmosphere, for example, contaminated people. He devised a method to destroy the microbes from milk by heating it at a specific temperature, a technique called to this day "pasteurization."

Another disease that was studied in the nineteenth century was cholera, a name derived from a Greek word—*kholera*—meaning diarrhea. An epidemic of cholera started in India in 1817 and spread through Europe. The English physician Dr. John Snow (1813–1858), analyzing in 1854 the concentration of cases in a small area of London, made the hypothesis that this fact was related to the contamination of the water from one of the pumps that supplied water to the population. This was confirmed, and later, Robert Koch (1843–1910), a German doctor, identified the cholera microbe. In 1882, Koch found the bacillus responsible for tuberculosis.

Alexander Fleming (1881–1955), a Scottish scientist working at St. Mary's Hospital Medical School in London, observed in 1928 that bacteria in a culture dish were killed by the growth of the mold *Penicillium notatum*. The mold spores had probably been transported by the wind. Fleming investigated this phenomenon and found that the mold was effective against microbes responsible for many diseases. The substance produced by the mold was called penicillin, and it became the first antibiotic. The result of Fleming's work was discovered by the Australian pathologist Howard Florey (1898–1968), who found how to purify penicillin by separating it from the liquid in the dish.

The Development of Medical Technology

The Englishman Robert Hooke (1635–1703) designed the compound microscope (i.e., with two lenses or more) in the seventeenth century. In 1830, an English physicist, Joseph J. Lister (1786–1869), made significant advances in the microscope. At the end of the century, in 1881, the Austrian physician Samuel Siegfried von Basch (1837–1905) invented the sphygmomanometer, a manometer to measure pulse pressure. An instrument to allow listening to the noises of the body was created by the French physician René Théophile

Laënnec (1781–1826) in 1816. He used a roll of paper with a cylindrical form to avoid the embarrassment of physical contact with a female patient. He perfected it in a second version and, using a wooden tube, invented the first stethoscope.

The contraction of the muscles is caused by electrical impulses. The human heart, for example, is a source of impulses that may be measured and studied to evaluate the organ's health. In 1903, at the beginning of the twentieth century, William Einthoven (1860–1927), a Dutch doctor born in Indonesia, working at the University of Leiden, published the details of the first instrument to study the electrical activity of the heart, the electrocardiograph; the result of the exam is an electrocardiogram (ECG).

Weaker electric waves are also associated with the activity of the brain, and therefore, their characterization became an essential tool in the study of this organ, with the instrument known as an electroencephalograph. The result of the application of this test is the electroencephalogram (or EEG). This method was described in 1875 by the English physician Richard Caton (1842–1926), who worked in Liverpool and studied electrical phenomena in the cerebral hemispheres of rabbits and monkeys.

As mentioned in Chap. 4, the electrical activity of the muscles, of the brain and the retina is measured with the techniques of electromyography (EMG), electroencephalography (EEG), and electroretinograms (ERM).

One crucial step in the development of medical technology was the first use of short-wavelength electromagnetic waves—like light waves, but of higher frequency—to generate images of the inside of the body. They were discovered in 1895 when Wilhelm Conrad Röntgen (1845–1923) [15] of the University of Würzburg was experimenting with a cathode ray tube[4] (See Chap. 6. The Secrets of Matter). In this device, a stream of electrons emitted by a filament was accelerated by an electric potential difference, and radiation was produced as it hit the tube wall. Röntgen noticed that a sample of a fluorescent substance outside the tube glowed when the stream of electrons was produced.

This unknown radiation was named "X-rays" by Röntgen, who studied its properties. X-rays incident on an object placed on a photographic plate produced an image, showing different degrees of transparency of the object, depending on the materials it contained. For example, X-rays incident on a

[4] He used cathode-ray tubes and Crookes tubes. The first device has a filament that is heated; the second device does not have a filament. In both types of tubes, electrons are accelerated by differences in voltage, and X-rays are emitted.

hand showed an image where the profile of the bones was perfectly visible. This illustrates the enormous importance of this discovery in medicine.

Another discovery that revealed similar radiation, this time emitted by a uranium salt, occurred when Henri Becquerel (1852–1908), Professor at the École Polytechnique in Paris, noticed that samples of a uranium salt kept in a drawer next to photographic plates left the plates sensitized.

In 1899, the Polish scientist Marie Curie (1867–1934), working at the University of Paris, announced the discovery of two new elements, polonium and radium, in samples of pitchblende, the mineral from which uranium was extracted. She created the name "radioactivity" to describe the phenomenon of the decay of nuclei with the emission of radiation. Later, it was demonstrated by Marie Curie's daughter, Irène Joliot-Curie (1926–1956), and her husband Frédéric Joliot-Curie (1900–1958) (both adopted the surname Curie) that radioactivity could be induced by irradiating otherwise stable nuclei. The resulting nuclei, called "radioisotopes," have found important medical applications. For example, radiation from the radioactive ^{60}Co isotope nuclei is used to treat cancer tumors. Radioactive isotopes are also used as markers in the study of metabolic processes.

The first electron microscope was made by the German physicist Ernst Ruska (1906–1988). An electron microscope is able to operate at much higher magnification than optical instruments. It uses magnetic fields that deviate the path of electrons, like lenses do to light rays.

Magnetic Resonance Imaging (MRI and fMRI)

In the 1970s, the power of nuclear magnetic resonance (NMR) (see Chap. 7) as a technique for the production of images was developed, beginning with the work of the American chemist Paul C. Lauterbur (1928–2007), then at the State University of New York [16], who reported the image of the cross-section of two vials containing water. This result, after many advances made in several different research laboratories, notably by Peter Mansfield (1933–2017) at Nottingham University, in England, led to this remarkable diagnostic tool, now known as magnetic resonance imaging, or MRI. MRI is based on the fact that, in principle, the intensity of the nuclear resonance signal is proportional to the number of resonating nuclei present in each region of a sample. By applying a magnetic field that varies from point to point, one obtains signals that arise from each small volume element of the sample. By showing on a screen the intensity of the resonance signal at each of these volumes within the sample, one can produce an NMR image.

Different images are obtained for each plane of interest that intersects the sample.

In the more common use of MRI, the resonating nuclei are protons, the nuclei of hydrogen atoms; the intensity of the signal from each region, therefore, measures the number of hydrogen nuclei. In the medical applications of MRI, the technique is applied to the human body, and the intensity of the signal shows the percentage of water (where most of the hydrogen atoms are located) in the soft human tissues, in the form of a computer image.

The magnetic resonance image is also sensitive to time evolution information. For example, after the application of the short pulses of radio waves, the time it takes for the resonance signals to die out can also be recorded, and this may vary from one kind of tissue to another, allowing a differentiation, for example, of the tissues of the brain: in the brain, these decay times are different for white matter and gray matter and for the cerebrospinal fluid [17]. The time evolution of the magnetic resonance signal is also sensitive to the resonant nuclei's motion; thus, the technique can be used to detect, for example, blood flow.

MRI is less invasive than X-rays computer tomography (CT, see below), since the radio waves do not present the severe health hazards associated with X-rays. However, the requirement of an intense magnetic field makes its use dangerous for patients with implanted pacemakers or metallic prostheses. Over a hundred million MRI diagnoses have been made, and there is no evidence of harmful effects of the highest magnetic fields used, in the 3–4 tesla range [18].

The possibility of recording images of human organs can be combined with the ability to furnish chemical information of the type provided by conventional NMR in a variation known as the functional MRI (fMRI) technique. This allows the visualization of the activity of different tissues, for example, of the brain, through changes in chemical composition or in blood flow. After injecting into the bloodstream a substance containing gadolinium atoms (the 'contrast medium'), the gadolinium magnetic moments shorten the relaxation rate of the protons, and consequently, the protons in the neighborhood will appear brighter [19]. This effect will make the tissue where it occurs stand out in the magnetic resonance image; this emphasizes the areas where the blood flow has increased. In the case of the brain, changes in neural activity are associated with changes in blood flow, and therefore, this effect provides a mechanism for mapping the brain areas that are more active in a given situation.

The standard MRI equipment uses a high magnetic field (~1 T). However, recent developments in technology have allowed operation at low fields (~0.1 T), characterized as "ultra-low-field" or "very-low-field." In 2020, the

US Food and Drug Administration approved a portable low-field scanner for use at 0.064 T for neuroimaging. It can be powered via a standard wall socket and does not require cryogenic liquids [20, 21]. This equipment requires the use of contrast agents.

Another important tool is Computerized Axial Tomography (CAT) (also known as Computer Assisted Tomography, or Computer Tomography), a technique [22] that generates X-ray cross-section images of the organ, e.g., the brain. An X-ray source emits short pulses of radiation, detected by an instrument that rotates around the head of the patient. A computer analyzes and integrates the X-ray data from several scans to construct cross-sectional images ("slices") of the organ being studied.

Blood-oxygen-level-dependent imaging, or BOLD-contrast imaging, is a method used in functional magnetic resonance imaging (fMRI) to observe different areas of the brain or other organs when they are activated. The physical basis of the BOLD method is the change in magnetic properties of the hemoglobin when oxygenated, discovered in the 1930s: the hemoglobin changes from paramagnetic deoxyhemoglobin to diamagnetic oxyhemoglobin.[5] A stimulus activates the neurons that draw deoxyhemoglobin, which induces a change observed in the MRI scanner. This method was introduced in 1990 [23, 24].

Gerhard Baule and Richard McFee, in 1966, first measured the magnetic field of the human body when they made the first magnetocardiogram (MCG) [25]. The magnetic field due to the activity of the heart is very small, in the range of $50–100 \times 10^{-12}$ T. This is much smaller than the Earth's magnetic field, of the order of 3×10^{-5} T. The detector usually employed to measure the MCG is the superconducting quantum interference device (SQUID) magnetometer. In 1974, the technique was used for the first time to monitor the heart of a fetus. The magnetocardiogram (MCG) and the electrocardiogram (ECG) may contain different information; thus, their use is complementary.

Magnetoencephalography (MEG) [26] is another technique for mapping brain activity, discovered in 1968, based on the detection of the magnetic fields generated by electrical currents that arise from the activity of the brain. These magnetic fields are typically a thousand times weaker than those measured in the heart and are also measured with SQUID magnetometers, installed in a helmet that is worn by the patient. SQUIDs constructed using high-temperature superconductors may use liquid nitrogen as a coolant, instead of liquid helium, which is more expensive and more difficult to handle.

---

[5] In an applied magnetic field, paramagnetic molecules develop a magnetic moment in the same direction as the field; diamagnetic molecules develop a weaker magnetic moment, opposite to the applied field.

Superconductors are materials that exhibit the phenomenon of superconductivity, i.e., their electrical conductivity is practically zero at low temperatures [27]. This effect was observed for the first time in 1911 by Heike Kamerlingh Onnes (1853–1926) in Leiden, Holland. This drop in resistance was first observed in mercury, as it reached a temperature of 4.2 degrees Kelvin (or – 269,3 C) [28].

Today, many other metals, alloys, and compounds are known to exhibit this behavior.

Besides the property of the striking reduction in conductivity, the superconductors also have a characteristic magnetic property: they behave as perfect diamagnets, i.e., as a sample becomes superconductive, the field lines originally existing are expelled. This is known as the Meissner Effect.[6]

A piece of iron near a magnet is magnetized in a direction opposite to that as the magnet. On the other hand, a superconductor is magnetized in the same direction of the magnet. Consequently, the piece of iron is attracted by the magnet, and the superconductor is repelled.

This force of repulsion may be sufficient to make one of the two objects float above the other.

This repulsion is the physical basis of the Maglev (from Magnetic Levitation) vehicles on rails that are presently operated in several countries. Levitating trains have some technological advantages over conventional trains, the most important one being the reduction of the losses arising from the friction of the wheels with the rails. One of these vehicles reached a speed of 603 km/h (375 mi/h) in 2015.[7] Superconductors have many other applications.

MEG and EEG have more temporal resolution than fMRI and PET (positron-emission tomography); the contrary is true for spatial resolution.

Other techniques that do not involve magnetic fields may also be used to study neural activity and the degree of oxygenation of the blood, for example, NIRS (Near Infrared Spectroscopy), employed in brain studies and in wrist-worn oxygenation meters [29]. This technique uses the fact that light in the near-infrared region penetrates more easily in the human tissue, since it is less absorbed by water and hemoglobin. The penetration of the light of this wavelength may distinguish oxygenated from deoxygenated hemoglobin.

[6] Discovered by the German physicist Walther Meissner (1882–1974) in 1933.

[7] Yao and Ma (2021).

Use of Magnetic Materials in Medicine

Hippocrates of Cos (c. 460–360 BC) used the magnetic oxides magnetite and hematite to control hemorrhages [30]. References to the use of magnetite may, in fact, describe applications of "magnesite," a magnesium carbonate with laxative properties.

In the *Natural History*, the encyclopedic work of the Roman naturalist Gaius Plinius Secundus, known as Pliny the Elder (23–79 AD), there appear some applications of magnetic minerals, but it is known that these inclusions were not critically filtered, reflecting both actual facts and superstitions.

The Greek Pedanius Dioscorides (c. 40–90 AD), born in Anazarbos and author of *De materia medica*, a 5-volume encyclopedia of medical matter (c. 64 AD), also mentions an external use of magnetite for "drawing out gross humors" [31]. Throughout history, many authors described medical applications of magnetic materials; in some cases, making use of the supposed laxative power of iron oxides.

The Indian system of medicine is called *Ayurveda*. It is described in two treatises, *Charaka Samhita* and *Susruta Samhita*, with Charaka and Susruta (or Sushruta) being the attributed authors, or writers that updated earlier texts, between 100 BC and 200 AD, in the first case, and 2000 BC–600 AD in the second text. The earliest description of a surgical use of the lodestone is found in the *Susruta Samhita*. In the Sucruta text, the magnet, which is in Sanskrit called "*Ayas Kanta*"—the "one loved by iron"—can be used to extract an iron arrow tip.[8]

In China, magnets were used to recover iron objects that were swallowed by children. This application is described in a publication of 1029.[9] A point of an iron knife accidentally swallowed was recovered with a magnet, requiring an incision along the throat, around the year 1635.[10] In the same century, a woman had accidentally swallowed an iron needle that remained in her throat for a total of 9 years. It was finally removed through a small incision, using a magnet in contact with her neck.

Small iron particles that adhere to the eyes have also been removed by using a magnet. In the nineteenth century, many ophthalmologists used magnets and electromagnets to remove metallic particles from the eyes.

A routine modern application of magnetism protects the cattle that eventually swallow while grazing barbs from barbed wire. Small magnets covered

[8] Häfely, U. (2007), Op. cit. p. 6.

[9] Chhieh Chung Fang of Chhien Wei-Yen of the year 1029 Needham, J [32].

[10] Häfely, U. (2007), Op. cit. p. 6.

with a plastic coating are placed in their stomachs and attract any iron fragments instead of allowing them to hurt the animal's vital organs.

During the past decades, the medical use of magnets has spread to fields as diverse as dentistry, cardiology, neurosurgery, oncology, and radiology, to mention only a few. The medical applications of magnets cover a very wide range of techniques. In surgery, magnet applications include the simple use of magnets in instrument holders placed on the patient's chest, to much more delicate functions, such as switching pacemakers or guiding tubes and catheters. Also, magnets may be used in the throat of patients to prevent sleep apnea.

Applications in vascular catheters that allow guidance from outside the body are used, both for intracranial encephalograms and to produce electrothrombosis. The latter is a blood clot inside a blood vessel that presents an inoperable arterial bulge (an aneurysm). Many applications of implanted magnets in the digestive tract and blood vessels are routinely used nowadays [33]. A list of current uses of magnets in medical applications, given in a recent work, includes connecting two small intestine segments, creation of anastomosis (i.e., union) between two blood vessels (e.g., in coronary artery bypass surgery), magnets placed in the anterior and posterior part of the urethra to avoid urinary incontinence, and so on[11].

Many experiments in biology use magnets; in a recent example, a study of eye motion of mice during sleep used a magnetic sensor and an implanted magnet beneath the conjunctiva of the eye [34].

Uses of Magnetism in Medicine: Magnetic Nanoparticles (MNPs)

A sample of magnetic material subjected to an oscillating magnetic field (a field that varies between +**B** and -**B)** changes its magnetization in a characteristic way, called its hysteresis curve. As the magnetization varies cyclically along this curve, the sample, in each cycle, gains magnetic energy and transforms this energy into thermal energy, i.e., it pumps thermal energy, liberating heat into the medium where it sits.

Magnetic nanoparticles (MNPs) are particles of magnetic material that have diameters in the range ($\sim 1–100 \times 10^{-9}$ m). For example, released in the body of a patient, in the neighborhood of a tumor, and submitted to oscillating magnetic fields, they liberate heat that may destroy it, a process favored by

[11] Lee et al. (2023), Op. cit. p. 4.

the fact that tumors are more sensitive to heat than healthy tissues. This process is called hyperthermia.

There are many other medical applications of these particles, including their use in diagnosis. Although formed of magnetic materials, the MNPs that are in the smaller size range do not behave as little magnets; they present what is called superparamagnetic behavior, i.e., they respond when a magnetic field is applied, but in the absence of a field they do not have stable magnetic moments, and consequently, they do not form clusters, a very important property for particles that are introduced into the body of a patient. Superparamagnetism is also a relevant phenomenon in magnetic recording (see Chap. 9).

Since they respond to an applied field, the particles can be guided by such a field (using a magnet, for example). Therefore, they may carry a medicine, delivering it directly to a tumor. The amount of medicine necessary in this form of delivery is much smaller than the dosage required via oral or injection with a hypodermic syringe.

These particles may be used not only in the treatment but also in the diagnosis of different diseases, such as cancer, infections, cardiovascular or autoimmune diseases, and neurological disorders [35, 36]. MNPs have also been used for pain management [37].

Another application of these particles is to increase the contrast in magnetic resonance images (MRI). Finally, as shown, the critical advantage of MNPs is their possibility of being manipulated by applying external magnetic fields.

The use of magnets in dentistry is another field of rapid expansion, mainly in the last decades [38]; this includes use in implants and orthodontic devices. In this field, magnetic particles have also been used for the targeted application of drugs, for the treatment of tumors, and for diagnosis. They have also been applied in the treatment of microflora-related infections.

Magnetic particles are also used to treat bone cancer, and the magnetic fields can stimulate the development of the bone-forming cells.

Other proposed uses of magnets: Mesmer and "animal magnetism".

In the eighteenth century, Friedrich Anton Mesmer (1734–1815), a German physician in Vienna, published *De planetarum influx in corpus humanum* ("The Influence of the Planets on the Human Body"*)* (1766), where he exposed his ideas on the influence of the planets on human health.

He later heard of experiments by English physicians with magnets and started to think in terms of "animal magnetism," a human faculty that could be restored by the therapeutic use of the magnet. Mesmer stroked with magnets the bodies of people and induced a state one would now describe as equivalent to a hypnotic trance. In 1776, he abandoned the use of the magnet but still referred to the phenomenon as animal magnetism. He believed that a "magnetic fluid" passed from him to his patients, inducing the trance.

Mesmer was a friend of Wolfgang Amadeus Mozart (1756–1791), who included a reference to mesmerism in one scene of the opera *Così fan tutte* (*Thus do they all*), of 1790. One of the characters pretends to be Mesmer and applies a large horseshoe magnet to two ladies, who recover immediately from their illnesses. Also, the first production of *Bastien und Bastienne*, a one-act *singspiel* (a type of musical work popular in the eighteenth century), was said to be held in Mesmer's gardens in 1768.

Mesmer could hypnotize people without the magnet, and he perfected his technique after he moved to Paris in 1778, adopting the *baquet*, a chest containing acid and equipped with iron parts said to be magnetized. The patients sat in a circle around the chest as Mesmer made passes over them. He claimed many cures, and in fact, he and his followers published hundreds of case histories; with time, his reputation soared. In Paris, he benefited from a general atmosphere favorable to science and, as a consequence, to therapies that appeared to be scientific [39]. It was said that he had been offered by the French government 20,000 francs to disclose his secrets, but he refused it.

The word "mesmerism" was employed for some time for the induction of this psychological state in the patients until it was substituted by "hypnotism," a term first used by James Braid (c. 1795–1860), a Scottish physician, in 1843.

Although hypnosis has evolved into a therapeutic tool of wide application [40], Mesmer's activities found growing opposition from French physicians. This fact led King Louis XVI to appoint in 1784 a commission to investigate Mesmer's practice, formed among others by Benjamin Franklin (1706–1790) and the French chemist Antoine-Laurent Lavoisier (1743–1794). The report of the panel was critical of the literal description of animal magnetism but supportive of the therapeutic "power of suggestion" [41].

Mesmer was finally accused of quackery and had to leave the country. His use of the *baquet* has survived in the circles of spiritualistic sitters, with no less scientific disrepute.

Mesmer's work had a part in the genealogy of modern psychoanalysis; the itinerary followed was mesmerism, hypnotism, and psychoanalysis, and the common concept has been the idea of transference, for Mesmer of a cosmic fluid, for Sigmund Freud of the creative sources of the unconscious [42].

References

1. Wightman, W. P. D. (1971). *The emergence of scientific medicine* (p. 6). Oliver & Boyd.
2. Lillie, M. C. (1998). Cranial surgery dates back to Mesolithic. *Nature, 391*, 854.
3. González-Darder, J. M. (2017). Cranial trepanation in primitive cultures. *Neurocirugia, 28*, 28–40.
4. Barnes and Noble, New York, and Taton R., Ed. (1966). *La Science Antique et Médiévale (des Origines a 1450),* 2e éd. révisée et mise à jour, PUF, Paris (1966).
5. Jouanna, J. (2012). *Greek medicine from Hippocrates to Galen, selected papers, chapter 1* (p. 16). Egyptian Medicine and Greek Medicine, Brill.
6. Sarton, G. (1927). *Introduction to the History of science* (Vol. 1, p. 96). From Homer to Omar Khayyam, Carnegie Institution of Washington, Baltimore.
7. Lindberg, D. C. (2007). *The beginnings of Western science* (2nd ed., p. 127). The University of Chicago Press.
8. McClellan, J. E., III, & Dorn, H. (2006). *Science and Technology in World History: An introduction* (2nd ed., p. 147). Johns Hopkins University Press.
9. Natarajan, K. (2008). Surgical instruments and endoscopes of Susruta, the sage surgeon of ancient India. *Indian Journal of Surgery, 70*, 219–223.
10. Sarton, G. (1927). *Introduction to the History of science* (Vol. 1, p. 708). From Homer to Omar Khayyam, Carnegie Institution of Washington, Baltimore.
11. Caillet, A. L. (1964). *Manuel Bibliographique des Sciences Psychiques ou Occultes.*, Tome II, Reprinted by B. de Graaf, Nieuwkoop (p. 198).
12. Boring, E. G. (1957). Hypnotism. In R. M. Elliott (Ed.), *A history of experimental psychology* (2nd ed., pp. 116–133). Prentice-Hall.
13. Wolf, A. (1962). *History of science, technology, and philosophy in the 16th and 17th century* (Vol. 1, 2nd ed., p. 298). George Allen & Unwin.
14. Porter, R. (Ed.). (1996). *The Cambridge illustrated history of medicine* (p. 159). Cambridge University Press.
15. Hurd, D. L., & Kipling, J. J. (1970). *The origins and growth of physical science* (Vol. 2, p. 324). Penguin Books.
16. Mourino, M. R. (1991). From Thales to Lauterbur, or from the lodestone to MR imaging: Magnetism and medicine. *Radiology, 180*, 593–612.
17. Bushberg, J. T., Seibert, J. A., Leidholdt, E. M., Jr., & Boone, J. M. (1994). *The essential physics of medical imaging* (p. 312). Williams & Wilkins.
18. Schenk, J. F. (2000). Safety of strong, static magnetic fields. *Journal of Magnetic Resonance Imaging, 12*, 2.
19. Islam, T., & Tsnobiladze, V. (2024). The application, safety, and recent developments of commonly used gadolinium-based contrast agents in MRI: A scoping review. *European Medicine of Journal, 9*(3), 63–73. https://doi.org/10.33590/emj/ZRVN2069

20. Oberdick, S. D., Jordanova, K. V., Lundstrom, J. T., Parigi, G., Poorman, M. E., Zabow, G. M., & Keenan, K. E. (2023). Iron oxide nanoparticles as positive T_1 contrast agents for low-field magnetic resonance imaging at 64 mT. *Science Reports, 13*, 11520.
21. Altaf, A., Shakir, M., Irshad, H. A., Atif, S., Kumari, U., Islam, O. W., Kimberly, W. T., Knopp, E., Truwit, C., Siddiqui, K., & Enam, S. A. (2024). Applications, limitations, and advancements of ultra-low-field magnetic resonance image: A scoping review. *Surgical Neurology International, 15*(218), 218.
22. Patel, P. R., & De Jesus, O. (2021). *CT scan*. Study Guide from StatPearls Publishing.
23. Ogawa, S., Lee, T. M., Kay, A. R., & Tank, D. W. (1990). Brain magnetic resonance imaging with contrast dependent on blood oxygenation. *Proceedings of the National Academy of Sciences of the United States of America, 87*, 9868.
24. Arthurs, O. J., & Boniface, S. (2002). How well do we understand the neural origins of the fMRI BOLD signal? *Trends in Neurosciences, 25*(1), 27–31. https://doi.org/10.1016/S0166-2236(00)01995-0
25. Roth, B. J. (2023). Biomagnetism: The first sixty years. *Sensors, 23*, 4218. https://doi.org/10.3390/s23094218
26. Fred, A. L., Kumar, S. N., Haridhas, A. K., Ghosh, S., Bhuvana, H. P., Sim, W. K. J., Vimalan, V., Givo, F. A. S., Jousmäki, V., Padmanabhan, P., & Gulyás, B. (2022). A brief introduction to magnetoencephalography (MEG) and its clinical applications. *Brain Sciences, 12*, 788. https://doi.org/10.3390/brainsci12060788
27. Yao, C., & Ma, Y. (2021). Superconducting materials: Challenges and opportunities for large-scale applications. *iScience, 24*, 102541.
28. van Delft, D. (2012). History and significance of the discovery of superconductivity by Kamerlingh Onnes in 1911. *Physica C, 479*, 30.
29. Sakudo, A. (2016). Near-infrared spectroscopy for medical applications: Current status and future perspectives. *Clinica Chimica Acta, 455*, 181–188.
30. Häfely, U. (2007). The history of magnetism in medicine. In W. Andrä & H. Nowak (Eds.), *Magnetism in medicine: A handbook* (2nd ed., p. 5). Wiley-VCH.
31. Duane, R. H. D. (1959). *The De Magnete of William Gilbert* (p. 27). Menno Hertzberger.
32. Needham, J. (2004). Science and civilisation in China. In *Physics and physical technology part I, physics* (Vol. 4, p. 236).
33. Lee, W. G., Evans, L. L., Johnson, S. M., & Woo, R. K. (2023). The evolving use of magnets in surgery: Biomedical considerations and a review of their current applications. *Bioengineering, 10*, 442.
34. Meng, Q., Tan, X., Jiang, C., Xiong, Y., Yan, B., & Zhang, J. (2021). Tracking eye movements during sleep in mice. *Frontiers in Neuroscience, 15*, 616760. https://doi.org/10.3389/fnins.2021.616760

35. Cardoso, V. F., Francesko, A., Ribeiro, C., Bañobre-López, M., Martins, P., & Lanceros-Mendez, S. (2018). Advances in magnetic nanoparticles for biomedical applications. *Advanced Healthcare Materials, 7*, 1700845.
36. Amorim, C. O. (2025). A compendium of magnetic nanoparticle essentials: A comprehensive guide for beginners and experts. *Pharmaceutics, 17*, 137.
37. Garello, F., Svenskaya, Y., Parakhonskiy, B., & Filippi, M. (2023). On the road to precision medicine: magnetic systems for tissue regeneration, drug delivery, imaging, and theranostics. *Pharmaceutics, 15*, 1812.
38. Yu, Y., & Li, X. (2024). Current application of magnetic materials in the dental field. *Magnetochemistry, 10*, 46.
39. Steptoe, A. (1986). Mozart, Mesmer and 'Cosi Fan Tutte'. *Music & Letters, 67*, 248–255.
40. Green, J. P., Laurence, J.-R., & Lynn, S. J. (2014). Hypnosis and psychotherapy: From Mesmer to mindfulness. *Psychology of Consciousness: Theory, Research, and Practice, 1*(2), 199–212.
41. Spiegel, D. (2012). Mesmer minus magic: Hypnosis and modern medicine. *International Journal of Clinical and Experimental Hypnosis, 50*(4), 397–406.
42. Schott, H. (1994). Neurogamies. De la relation entre mesmerism, hypnose et psychanalyse. In J. Clair (Ed.), *L'Âme au Corps, Arts et Sciences 1793–1993* (pp. 142–153). Galimard.

Further Reading

Häfely, U. (2007). The History of magnetism in medicine. In W. Andrä & H. Nowak (Eds.), *Magnetism in medicine: A handbook* (2nd ed.). Wiley-VCH.

Jouanna, J. (2012). *Greek medicine from Hippocrates to Galen, selected papers, chapter 1*. Egyptian Medicine and Greek Medicine, Brill.

Lee, W. G., Evans, L. L., Johnson, S. M., & Woo, R. K. (2023). The evolving use of magnets in surgery: Biomedical considerations and a review of their current applications. *Bioengineering, 10*, 442.

McClellan, J. E., III, & Dorn, H. (2006). *Science and Technology in World History: An introduction* (2nd ed.). Johns Hopkins University Press.

Merrill, R. T., McElhinny, M. W., & McFadden, P. L. (1998). *The magnetic field of the earth*. Academic Press.

Mielczarek, E. V., & McGrayne, S. B. (2000). *Iron*. Nature's Universal Element.

Needham, J. (2004). Science and civilisation in China. In N. Sivin (Ed.), *Part VI: Medicine* (Vol. 6). Cambridge University Press.

Sambursky, S. (1974). *Physical thought: From the Presocratics to the quantum physicists*. Pica Press.

Sarton, G. (1927). *Introduction to the History of science* (Vol. 1). Robert E. Krieger Publishing Company, reprinted (1975).

Sutcliffe, J., & Duin, N. (1992). *A History of medicine, from prehistory to the year 2020*. Barnes & Noble.

Taton, R. (1966). Histoire générale des sciences. Tome I: La science antique et médiévale (des origines à 1450). In *révisée et mise à jour* (2nd ed.). PUF.

Wolf, A. (1962). *History of science, technology, and philosophy in the 16th and seventeenth century* (Vol. 1, 2nd ed.). George Allen & Unwin.

9

Magnets that Remember: A Pile Higher than the Earth-Moon Distance

Memory is the mother of all wisdom.
(Attributed to the Greek tragedian Aeschylus (c. 525–c. 455 BC)).

Summary The written word revolutionized the diffusion and accumulation of knowledge, leading to the invention of the book. The text of an average book contains about one million bytes of information (1 MB). In a magnetic memory, data are recorded as tiny magnetized regions on a recording medium, in magnetic hard disks (HDDs), tapes, and magnetic stripe cards. In the first magnetic tape system (1951), the data density was about 100 million times smaller than today's. The data density in HDDs has increased almost a thousand million times since 1956. HDDs have recently incorporated energy-assisted magnetic recording (EAMR); also, new HDDs may use superposed tracks. The total information produced worldwide per year will reach about 2×10^{24} bytes in 2035. A pile of books containing this amount of information will be 40,000,000,000 thousand km high! This total illustrates the information explosion, one of the characteristic aspects of our era. In 2025, the information contained in all the data centers where information is stored reaches 2×10^{23} bytes. Magnetic storage is, therefore, the primary tool for preserving humankind's memory.

In the novel cycle *Remembrance of Things Past*, the masterpiece of French writer Marcel Proust (1871–1922), there is a famous passage [1] inspired by an episode experienced by the author. In it, the central character describes how the smell of a cake—a *madeleine*—evokes a chain of images connected to

A. P. Guimarães, *A Longstanding Attraction*, https://doi.org/10.1007/978-3-032-02006-2_9

his youth: "And as soon as I had recognised the taste of the piece of *madeleine* (...) immediately the old gray house upon the street, where her room [his aunt's] was, rose up like a stage (...) and with the house, the town, from morning to night and in all weathers, the Square where I used to be sent before lunch, the streets along which I used to run errands, the country roads we took when it was fine."

In the example of Proust's novel, information stored in the brain is evoked many years later by an external stimulus, the smell of a *madeleine*; other recollections may arise from an image, a name, or a sound. Despite the brain's ability to store a prodigious amount of information, human memory may fail, and a secure means is required to record data and transmit knowledge. Humankind invented writing in response to this need; this great invention first appeared among the Sumerians in Mesopotamia in the fourth millennium before our era, as mentioned in Chap. 2.

The written word created a revolution in the diffusion and accumulation of knowledge. It led to the invention of the book, a long time after the period in which clay tablets or papyrus rolls were used. The library of the Museum of Alexandria, one of the most celebrated institutions of the ancient world, was founded in the third century BC and destroyed over the following centuries. At the height of its existence, it contained scrolls corresponding to some 50,000 average books [2],[1] a notably rich collection for the period (some authors argue that there is much uncertainty in these numbers)[2].

The first volumes of pages joined on the sides, as in present-day books, are the *codex* (from Latin *caudex*, the trunk of a tree), substituting the scrolls from the fourth century AD.

The invention of printing in the fifteenth century was a turning point of enormous importance. The English scientist and statesman Francis Bacon (1561–1626) wrote [4]: "Again, we should notice the force, effect, and consequences of inventions, which are nowhere more conspicuous than in those three which were unknown to the ancients; namely, printing, gunpowder, and the compass. For these three have changed the appearance and state of the whole world…"

Under the invention of printing, one describes the invention by Johannes Gutenberg (c. 1400–1468) in Mainz in the late 1430s of movable type, a form of printing that used an arrangement of individual characters under pressure against paper pages. This invention had no connection to earlier developments in printing in Asia; the Chinese invented movable type around

[1] Ancient numbers on the size of the Library of Alexandria "do not deserve any credence…," Bagnall [3].

[2] Bagnall, S. (2002), Op. cit.

Table 9.1 Number of bytes and units

Unit	Size	Abbreviation
One kilobyte	10^3 B	1 kB
One megabyte	10^6 B	1 MB
One gigabyte	10^9 B	1 GB
One terabyte	10^{12} B	1 TB
One petabyte	10^{15} B	1 PB
One exabyte	10^{18} B	1 EB
One zettabyte	10^{21} B	1 ZB
One yottabyte	10^{24} B	1 YB

1040. The same technology was developed in Korea in 1403; in this same year, it was published in China, an encyclopedia with 937 volumes [5].

The new technology spread rapidly, and by 1500, some 13,000 works were printed. One can compare this number to the estimated number of titles of books presently published worldwide each year, some four million; a total of some 2.2 billion books are sold per year.[3]

How can one quantify the amount of information contained in a book? To specify one text character (letter, space, or punctuation mark), one needs eight units of data, or 'bits' (from 'binary digit'), each representing a "0" or a "1." A set of eight bits corresponds to one 'byte.' An average book without illustrations is estimated to contain about one million bytes of information, or one megabyte (MB) (see units in Tables 9.1 and 9.2).

Contrasting with the Alexandria library, a modern national library, such as the US Library of Congress, has some twenty-four million books on its shelves.[4, 5] They contain, in the text of the volumes alone, some twenty-four million million bytes, or twenty-four trillion bytes (or 24 terabytes, i.e., 24×10^{12} bytes, a number of bytes given by the number 24 followed by twelve zeros).[6]

This information outburst has demanded new forms of information storage; the discovery of magnetic recording afforded a practical means of preserving and making readily available sound, text, numeric data, and images.

---

[3] https://wordsrated.com/book-sales-statistics.

[4] More than 25.77 million cataloged books in the Library of Congress classification system, more than 15.99 million items in the non-classified print collections, including books in large type and raised characters, incunabula (books printed before 1501), monographs and serials, music, bound newspapers, pamphlets, technical reports, and other printed material; more than 138.5 million items in the non-classified collections. https://www.loc.gov/about/general-information/#year-at-a-glance.

[5] Morrison, P. & Morrison, P. (1998). op. cit., p. 115.

[6] https://physics.nist.gov/cuu/Units/binary.html.

Table 9.2 Some estimated quantities of information

A written page	5 kB = 5 × 10^3 bytes
Complete works of Shakespeare	5 MB = 5 × 10^6 bytes
US congress library (only texts of books)	24 TB = 24 × 10^{12}bytes
Text of the holy bible	5 MB = 5 × 10^6 bytes
Estimated number of words ever spoken	10^{19}
Information produced in 2023	120 ZB = 1.2 × 10^{24} bytes
Information produced in 2035 (predicted)	2100 ZB = 2.1 × 10^{24} bytes = 2.1 yottabytes
Information in data centers (2024)	100 ZB = 10^{23} bytes

To illustrate the basic idea behind magnetic recording, consider lava flowing from a volcano. Before the lava solidifies, the Earth's magnetic field orients those atoms with magnetic moments (i.e., atoms that are small magnets), and a magnetization, called in this case 'thermoremanent magnetization,' results. If one now brings close to the rock a device able to detect this change, the "written" rock magnetization can be "read." This analogy illustrates the principle behind magnetic recording.

In a magnetic recording apparatus, data are recorded when electrical impulses are translated into a pattern of tiny magnetized regions on a recording medium. For example, a film of magnetic material is deposited onto a tape or a disk. First, an electric signal is amplified; it then generates a magnetic field that is applied to the medium, creating the required magnetization. This magnetization is read by reversing the process: the magnetic field produced by the medium generates an electric signal related to the magnetization, which is detected when a reader or sensor moves past it.

Magnetic recording was invented at the end of the nineteenth century by the Danish engineer Valdemar Poulsen (1869–1942), who then worked at the Copenhagen Telephone Company (KTAS). In the Copenhagen Company, as in several others worldwide at that time, there was an interest in devising a system for recording telephone messages. This demand stimulated Poulsen to experiment with magnetic recording, trying initially with an electromagnet that magnetized a circular saw [6]. He only succeeded in his attempts when he tried using as a magnetic medium a 1-m-long steel wire held by two nails on a piece of wood. He moved along the wire an electromagnet connected to a microphone. The electrical signal generated by the microphone magnetized the wire, recording his voice.

Poulsen perfected his idea and filed the first patent in August 1898 for a recorder that used a drum to wind the steel wire, the 'Telegraphone.' He left the telephone company shortly after and started to work on developing his projects in association with the engineer and friend Peder Olaf Pedersen

(1874–1941). The invention provoked much interest at the time, and Poulsen was awarded [7] the grand prize at the Paris Exhibition of 1900. In the same exhibition, the oldest extant voice recording, of Emperor Franz Joseph of Austria (1830–1916), was made.

Credit cards and automated teller machine (ATM) cards also use a magnetic stripe where information on the owner's account, card number, and, in some cases, their personal identity number (PIN) are stored. In this case, magnetic technology competes with cards with embedded chips (smart cards), which were introduced later. Stripe cards hold information that stays the same. Cards with chips carry a digital code that changes after each operation by the owner, e.g., a purchase or a bank transfer. The latest generation of magnetic cards with chips is that of the EMV cards (from Europay, Mastercard, and Visa).

One also finds magnetic stripes on some subway tickets, plane boarding cards, and other cards and badges.

Magnetic tape is a form of magnetic storage that was first used in the UNIVAC I computer in 1951. The tape is an essentially bidimensional medium, with a thin layer of magnetic material deposited on a plastic. Magnetic tapes have sequential storage access, i.e., to read data in a specific region in the tape, a length of tape is wound on the reel, bringing that region to the read/write head. Present-day magnetic tapes exhibit a much larger capacity than the UNIVAC 1 system: 30 TB compared to 184 kB, an increase of about 100 million times.

The enormous expansion in the use of digital computers since the introduction of the first personal computers in the late 1970s was one of the factors that spurred the improvement of magnetic recording systems for data storage.

Earlier, small ferrite rings were used as magnetic memory elements in IBM computers, beginning in 1955 [8].

Hard disks, the familiar computer magnetic hard disk drives (HDDs), were introduced by IBM in 1956, with data stored at an areal density [9] (AD) of 2 kbits/inch^2 (or 3 bits/mm^2) (Fig. 9.1a).

Unlike magnetic tapes, hard drives provide direct access, i.e., any point in the disk can be reached in a very short time.

Computer data are stored in digital form, i.e., as a sequence of '0's and '1's, corresponding to regions magnetized in opposite directions, in the case of the hard disks, perpendicular to their plane.[7]

Hard disks have a two-dimensional recording medium, a film containing cobalt-based magnetic particles deposited onto an aluminum alloy or glass

[7] Before 2005, the disks were magnetized tangentially to their surface.

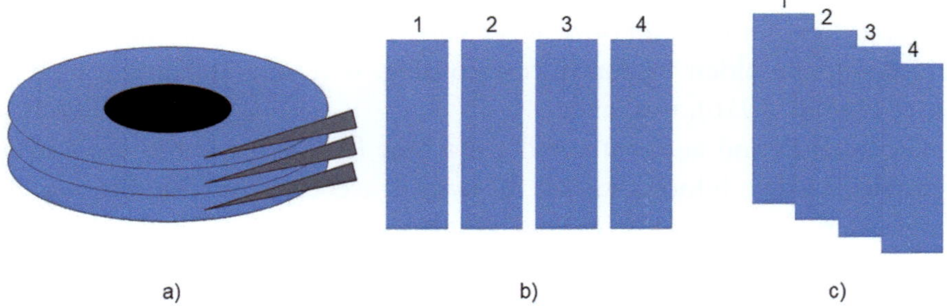

Fig. 9.1 **(a)** Schematic view of an HDD drive, showing three platters and the corresponding read-write heads; **(b)** Conventional layout of the recording tracks on a magnetic hard disk; **(c)** Layout of the tracks in shingled magnetic recording (SMR)

disk. In the HDD, a read-write head moves on an air cushion at a minimal distance of a few nanometers (millionths of a meter) from the disk's surface, which turns thousands of times per minute. The write head records each bit in a minimal area, presently with a demonstrated areal density (AD) of 2.77 Tbits/inch^2 (or 2.77×10^{12} bits per square inch).

In the new disk units, the 'read' head senses the magnetic field due to the magnetized disk through the change that it produces in the resistance of the material of the head. This magnetoresistance effect is the basis of the read heads called spin valves, introduced in the early 1990s. They are the most sensitive sensors,[8] formed by a structure with several layers, which vary the perpendicular resistivity under the influence of the magnetic field due to the magnetic media.

Magnetic recording units for computers (HDDs) have evolved at a fantastic speed. Compare the magnetic disks in 1956 with an areal density of 2 kB per square inch [10] with the present magnetic disk area density of about 1 TB per square inch[9] (or approximately 1.5 GB/mm^2), an increase by a factor of 500 million! From 1980 to 2010, the cost per data bit was reduced by a factor of about 1,000,000.

The magnetic layer in an HD consists of magnetic grains; each bit is recorded in a group of neighbor grains. Smaller and smaller grains have to be used to increase magnetic data density. However, as their size is reduced, the magnetization of individual grains becomes unstable, a phenomenon called superparamagnetism (SPM). One can avoid superparamagnetism by using materials with a higher magnetic anisotropy, i.e., materials requiring higher

[8] Albuquerque, G. et al., (2022). op. cit.

[9] Albuquerque, G. et al., (2022). op. cit.

magnetic applied fields to turn their magnetization direction. With such a type of material, the magnetic fields produced by the recording heads, required to change a "0" bit to a "1" bit, or vice versa, have now to be more intense.

This example of competing effects, in this case, using smaller grains to increase the information density on the hard drive, decreases the time stability of the recorded information, which, to be compensated, requires materials with increased magnetic anisotropy. This is an instance of the so-called magnetic recording trilemma, i.e., a dilemma with three options where the demands to advance magnetic recording are conflicting.

The three vertices of the trilemma correspond to three desired properties of the magnetic media:

(a) To favor writability, the magnetic material must have low magnetic anisotropy, or else the recording head would need to apply high magnetic fields.
(b) To favor high signal-to-noise ratios, smaller grains would be preferable;
(c) To assure thermal stability, larger grains, and high-anisotropy materials would be favored.

The magnetic recording and reading processes have recently incorporated new technology generally known as energy-assisted magnetic recording (EAMR), which comprises heat-assisted magnetic recording (HAMR) and microwave-assisted magnetic recording (MAMR).

In heat-assisted magnetic recording (HAMR) technology, the need to change to higher recording magnetic fields is avoided by locally heating the grains immediately before the recording head passes above them. A laser pulse is used to heat the grains; this is effective because magnetic anisotropy decreases with increasing temperature.

In the alternative technology, microwave-assisted magnetic recording (MAMR), an oscillator induces the precession of the magnetization of the HDD's magnetic grains. Once the magnetizations begin to precess, they are more easily turned by a magnetic field, demanding smaller applied fields by the recording head.

The storage density in hard disks has also evolved by implementing different layouts of the magnetic tracks where the data are recorded. In conventional magnetic recording, the individual tracks do not overlap (Fig. 9.1b). However, other recently adopted layouts allow the superposition of the tracks.

In one of the technologies that has already been implemented—shingled magnetic recording (SMR)[10]—the tracks are partially superposed, and a set of tracks forms a band (Fig. 9.1c). The information necessary to deal with updating tracks may be obtained using either an array of closely spaced reader heads or by storing information in the memory (which requires more complex electronics) and making several passes of a single head. Different forms of superposition are used in interlaced magnetic recording (IMR) and two-dimensional magnetic recording (TDMR).

According to the report *Rethink Data*[11] prepared by the International Data Corporation (IDC), it is estimated that the amount of new data created in the year 2025 reaches 175.8 ZB, compared to 18.2 ZB in 2015, an increase of a factor of ten in ten years. This increase illustrates the information explosion phenomenon, one of our era's characteristic aspects.

One can get an idea of how significant this enormous amount of data is by comparing it to the data contained in a single book. We have mentioned that a book has the order of one million bytes (10^6 bytes, or 1 MB) of information. Based on the report mentioned above, for 2025, the new data created in 1 h will be given by 175×10^{21} divided by the number of hours in the year (8760 h), i.e., 2×10^{19} bytes/hour. Assuming that the books in a library have an average thickness of 2 cm (0.02 m) and each has 10^6 bytes of information, the height of this pile with the information produced in an hour is approximately 400,000,000 km high, or 1000 times the Earth-Moon distance![12]

Many companies and individuals use cloud services: these virtual resources are available to store data and perform operations with it. These services require physical facilities called data centers, where stored data is estimated to have increased tenfold between 2018 and 2025.[13] Magnetic storage is the main form of storing information in these centers through the use of magnetic hard disks (HDDs) and magnetic tape; most data were stored on spinning disks by 2024, and 18% on magnetic tape, according to IDC's estimate.[14]

[10] So named for the similarity with overlapping rows as a covering for roofs.

[11] Rethink Data, *Seagate Technology Report*, Seagate Corporation https://www.seagate.com/content/dam/seagate/migrated-assets/www-content/our-story/rethink-data/file:s/Rethink_Data_Report_2020.pdf.

[12] A pile of such average books containing a total of 2×10^{19} bytes would have $2 \times 10^{19}/10^6 = 2 \times 10^{13}$ books. The height of this pile will (assuming a thickness of 2 cm), therefore, be $2 \times 10^{13} \times 0.02 = 4 \times 10^{11}$ m, or 4×10^8 km = 400,000,000 km high. The average Earth-Moon distance is 382,500 km; therefore, $4 \times 10^8/3.825 \times 10^5$ = approximately 10^3. Therefore, the pile would be one thousand times this distance!

[13] https://www.fortinet.com/resources/cyberglossary/data-center.

[14] The Future of Data Center Storage, *Horizon Editorial.* International Data Corporation, February 15, (2022).

Other forms of storage, based on semiconductor devices, such as solid-state drives (SSDs), are increasing their share of the data; the basis of the data centers, however, consists of hard disk drives (HDDs) and magnetic tape drives.

In 2028, the total amount of information contained in all the data centers in the world will reach 394 zettabytes (394×10^{23} bytes) [11].

Computer programs using artificial intelligence (AI) need large volumes of data for training. This fact is likely to increase the demands on the data centers.[15] AI may also impact magnetic memory technology by supporting the design of new materials—compounds and alloys—employing more abundant elements. Data-driven search for materials, including magnetic alloys and compounds, has superseded the trial-and-error approach, becoming the paradigm of new materials design [12]. These searches use many available databases organized according to structural properties, thermodynamic properties, electronic structure, cohesive energies, organic and inorganic compounds, etc. A list with more than 50 databases is illustrated in the previous reference.

The increase in the total data created in recent decades is also accounted for by the Internet of Things (IoT) growth. IoT describes the grid of interconnected devices installed in factories, homes, shops, vehicles, public spaces, and services [13]. These devices have sensors and are interconnected via Internet, through public or private grids. This growing network has already increased the productivity of the industries and agriculture. Healthcare, both in hospitals and remotely, has also benefited. The yearly output of data generated by IoT (Internet of Things) is expected to be multiplied by a factor of three between 2020 and 2030 [14].

In 2010, data centers already used 1% of the global electricity. It is estimated that they will use around 3–13% in 2030 [15]. The same work predicts that, in the worst-case scenario, the operation and production of consumer devices, communication networks, and data centers may use 2030 as much as 51% of global electricity. Global data center use of electricity in 2021 was 220–320 TWh, or around 0.9–1.3% of global final electricity demand.

The data center total does not include energy used for cryptocurrency mining, which was 100–140 TWh in 2021, according to the International Energy Agency (IEA).[16]

Another concern related to the growth of data centers is their impact on the environment; the carbon footprint of data centers corresponded, in 2021, to

[15] https://www.schroders.com/en-us/us/individual/insights/how-ai-is-set-to-accelerate-demand-for-data-centres/.

[16] https://www.iea.org/reports/data-centres-and-data-transmission-networks.

0.5% of total greenhouse emissions in the US [16]. This footprint is greater than that of the airline industry.

The rapid progress in magnetic storage technology has led, in several decades, to an increasing volume of the information in magnetic form. This amount is much more significant than the total information in the approximately one million new books published worldwide every year.

Magnetic storage is, therefore, the main form of recording the information produced by individuals, corporations, banks, the administration of all countries, and scientific laboratories.

Magnetic recording technology has attained an extraordinary position in our time, being responsible for the custody of most of the information or knowledge generated worldwide.

The wonder about the attraction of the lodestone has taken us a long way, from a curious phenomenon that was the object of awe and bafflement to an aspect of Nature present in every scale of complexity, from subatomic particles to galaxies.

This road leads from the first practical device using magnetic forces—the compass—to powerful tools for investigating the innermost secrets of the human body or to novel and fantastic ways of dealing with information, from the early compass that guaranteed the safety of sailors to the protection of humankind's wealth of knowledge.

References

1. Proust, M. (1982). Remembrance of things past. In *Swan's Way.*, translated by Moncrieff, C. K. S. & Kilmartin, T. (Vol. 1, p. 51). Vintage Books.
2. Morrison, P., & Morrison, P. (1998). The sum of human knowledge? *Scientific American, 279*, 115.
3. Bagnall, S. (2002). Alexandria: Library of Dreams. *Proceedings of the American Philosophical Society, 146*(4), 348.
4. Bacon, F. (1911). Aphorism 129. In J. Devey (Ed.), *Novum Organum.* P. F. Collier & Son.
5. McClellan, J. E., III, & Dorn, H. (2006). *Science and Technology in World History: An introduction* (2nd ed., p. 125). Johns Hopkins University Press.
6. Jorgensen, F. (1999). The inventor Valdemar Poulsen. *Journal of Magnetism and Magnetic Materials, 1*, 193.
7. Stevens, K. W. H. (1995). Magnetism. In L. M. Brown, A. Pais, & B. Pippard (Eds.), *Twentieth century physics* (Vol. II, p. 1158). Institute of Physics Publishing.
8. Bashe, C. J., Johnson, L. R., Palmer, J. H., & Pugh, E. W. (1989). *IBM'S early computers* (p. 239). MIT Press.

9. Wang, S. X., & Taratorin, A. M. (1999). *Magnetic information storage technology* (p. 10). Academic Press.
10. Piramanayagam, S. N., & Chong, T. C. (2012). *Developments in Data storage: Materials perspective*. Wiley.
11. Morgan, S., The 2020 Data Attack Surface Report. *Cybercrime Magazine*. https://cybersecurityventures.com/wp-content/uploads/2020/12/ArcserveDataReport2020.pdf
12. Singh, P., Del Rose, T., Palasyuk, A., & Mudryk, Y. (2023). Physics-informed machine-learning prediction of Curie temperatures and its promise for guiding the discovery of functional magnetic materials. *Chemistry of Materials, 35*, 6304–6312.
13. Gillis, A. S. (2023). *What is the Internet of Things (IoT)?* 01 Aug. https://www.techtarget.com/iotagenda/definition/Internet-of-Things-IoT
14. Mansour, M., Gamal, A., Ahmed, I. A., Said, L. A., Elbaz, A., Herencsar, N., & Soltan, A. (2023). Internet of things: A comprehensive overview on protocols, architectures, technologies, simulation tools, and future directions. *Energies, 16*, 3465. https://doi.org/10.3390/en16083465
15. Andrae, A. S. G., & Edler, T. (2015). On global electricity usage of communication technology: Trends to 2030. *Challenges, 6*, 117–157.
16. Siddik, M. A. B., Shehabi, A., & Marston, L. (2021). The environmental footprint of data centers in the United States. *Environmental Research Letters, 16*, 064017.

Further Reading

Albuquerque, G., Hernandez, S., Kief, M. T., Mauri, D., & Wang, L. (2022). HDD reader technology roadmap to an areal density of 4 Tbpsi and beyond. *IEEE Transactions on Magnetics, 58*, 3100410.

Bhushan, B. (2018). Historical evolution of magnetic data storage devices and related conferences. *Microsystem Technologies, 24*, 4423–4436. https://doi.org/10.1007/s00542-018-4133-6

Krishnan, K. K. (2016). *Fundamentals and applications of magnetic materials*. Oxford University Press.

McClellan, J. E., III, & Dorn, H. (2006). *Science and Technology in World History: An introduction* (2nd ed.). Johns Hopkins University Press.

Piramanayagam, S. N., & Chong, T. C. (2012). *Developments in Data storage: Materials perspective*. John Wiley.

Vedmedenko, E. Y., Kawakami, R. K., Sheka, D. D., Gambardella, P., Kirilyuk, A., Hirohata, A., Binek, C., Chubykalo-Fesenko, O., Sanvito, S., Kirby, B. J., Grollier, J., Everschor-Sitte, K., Kampfrath, T., You, C.-Y., & Berger, A. (2020). The 2020 magnetism roadmap. *Journal of Physics D: Applied Physics, 53*, 453001.

Timeline

2,000,000 BC

- Stones shaped for use as tools.

3,500 BC

- Rise of cities in the Tigris–Euphrates delta, the Indus, and Nile valleys.
- c. 3,000 BC: origins of the Ayurveda, the Indian medicine system.

c. 2000 BC

- Use of copper, later bronze.

2000 BC

- Early use of iron.
- 2000-1500 BC: Mention of "grasping hematite" for magnetite, in a list of commodities, Mesopotamia.

A. P. Guimarães, *A Longstanding Attraction*, https://doi.org/10.1007/978-3-032-02006-2

1000 BC

- First half of the first millennium BC: "grasping hematite" in a lexical text, Mesopotamia.
- Sixth century BC: Thales of Miletus (c. 640 BC-546 BC), in Greece, according to Aristotle (384-322 BC), considered that the magnet had a soul.

500 BC

- (c. 479 – c. 300 BC): Chinese medicine text *Nei Ching* (*Yellow Emperor's Classic of Internal Medicine*).
- The Greek Hippocrates of Cos (c. 450–c. 370 BC), known as the "father of Medicine. "
- 240 BC: In China, the book *Lü Shih Chhun Chhiu* ("*Master Lü's Spring and Autumn Annals*") describes the properties of the magnet.
- First century BC: Lucretius (95 BC-55 BC): The name magnet comes from the province of Magnesia; attraction is due to a stream of particles emitted by the magnet.
- Galen of Pergamon (129–216 AD), Greek physician.

500 AD

- Eighth or ninth century: Magnetic declination (the angle between the direction of a meridian and the direction of the magnetic field on a given point on the Earth) was known in China.
- Avicenna (or Ibn Sina) (980-1037), published the *Canon Medicina.*

1000 AD

- 1044: *Wu Ching Tsung Yao* ("*Collection of the Most Important Military Techniques*") was published, with a description of the compass.
- 1150: Probable date of appearance of the compass in Europe.
- 1269: Pierre de Maricourt (Petrus Peregrinus) (born *circa* 1220) publishes *the Epistola de Magnete* ("Letter on the Magnet"), in which he summarizes the current knowledge on magnetism.
- 1455: Johannes Gutenberg (c. 1400-1468) prints the 42-line Bible.

1500 AD

- 1543: Nicolaus Copernicus (1473-1543) publishes *De revolutionibus orbium coelestium* ("On the Revolutions of the Celestial Spheres").
- 1558: Giovanni Battista (or Giambattista) della Porta (1535-1615) publishes *Magia Naturalis* ("Natural Magic"), with many references to magnetism.
- 1581: Robert Norman (fl. 1590) publishes the Newe Attractive (1581).
- 1590: The microscope was invented.

1600 AD

- 1600: William Gilbert (1544-1603) publishes *De Magnete*, where he states that the Earth is a magnet.
- 1610: Galileo Galilei (1564-1642) publishes *Sidereus Nuncius* ("The Starry Messenger").
- 1618-1621: Johannes Kepler (1571-1630) publishes his *Epitome Astronomiae Copernicanae* ("Epitome of Copernican Astronomy"), where he compares the attraction of the Sun with that of the magnet.
- 1641: Athanasius Kircher (1601-1680) publishes M*agnes, sives De Arte Magnetica* ("The Magnet, or About the Magnetic Art").
- 1644: René Descartes (1596-1650) publishes the *Principles of Philosophy*, where he ascribes the magnetic attraction to a flow of 'threaded particles.'
- 1676: The compass in a ship is affected after being hit by lightning.
- 1687: Isaac Newton (1642-1727) publishes the *Principia Mathematica.*

1700 AD

- 1701: Edmund Halley (1656-1742) publishes the first geomagnetic chart.
- 1745-1746: Georg von Kleist (c. 1700-1748) and Pieter van Musschenbroek (1692-1761) invented the Leyden jar.
- 1752: Benjamin Franklin (1706-1790) performs experiments with atmospheric electricity.
- 1785: Charles Augustin de Coulomb (1736-1806) reports that the magnetic force falls with the square of the distance.
- 1780s: Luigi Galvani (1737-1798) obtains the first indications that chemical reactions may induce the flow of electricity.

1800 AD

- 1800: Alessandro Volta (1745-1827) publishes his discovery of the electric battery.
- 1816: René Théophile Laënnec (1781-1826) invents the stethoscope.
- 1820: Hans Christian Oersted (1777-1851) discovers that electric currents produce a magnetic field.
- 1820-21: André-Marie Ampère (1775-1836) attributes the magnetism of matter to 'molecular' currents.
- 1831: Michael Faraday (1791-1867) discovers electromagnetic induction, the phenomenon whereby a variable magnetic field induces an electric current.
- 1845: Michael Faraday (1791-1867) discovers diamagnetism and paramagnetism.
- 1864: James Clerk Maxwell (1831-1879) publishes "A Dynamical Theory of the Electromagnetic Field."
- 1869: Dmitri Mendeleev (1834-1909) presents the Periodic Table, with 63 elements.
- 1885-1889: Heinrich Hertz (1857-1894) detects radio waves and identifies their electromagnetic character.
- 1890: James Alfred Ewing (1855-1935) discovers hysteresis, the phenomenon responsible for the characteristic shape of the magnetization curve versus the applied magnetic field of a magnetic sample.
- 1895: Pierre Curie (1859-1906) discovers the law of variation of the magnetism of paramagnetic materials with temperature (known as Curie Law).
- 1895: Wilhelm Conrad Röntgen (1845-1923) discovers X-rays.
- 1897: Joseph John Thomson (1856-1940) discovers the electron.
- 1898: Valdemar Poulsen (1869-1942) invents magnetic recording.

1900 AD

- 1900: Max Planck (1858-1947) introduces the idea of the quantum.
- 1905: Albert Einstein (1879-1955) publishes the special relativity theory.
- 1905-1910: Paul Langevin (1872-1946) publishes a theory of paramagnetism, the phenomenon whereby a nonmagnetic material, under the influence of an applied magnetic field, exhibits a significant magnetization.
- 1909: Ernest Rutherford (1871-1937): Experiment with alpha particles that determines how small the nuclei are.

- 1911: Heike Kamerlingh Onnes (1853-1926) discovers superconductivity.
- 1913: Niels Bohr (1885-1962) publishes a model for the atom.
- 1916: Albert Einstein (1879-1955) publishes the general relativity theory.
- 1919: Joseph Larmor (1857-1942) proposes the self-exciting dynamo theory to explain the magnetism of the Sun.
- 1922: Otto Stern (1888-1969) and Walter Gerlach (1889 - 1979) demonstrated that the spatial orientation of the angular momentum is quantized.
- 1922-1924: Louis de Broglie (1892-1987) proposes the hypothesis that ordinary matter exhibits wavelike behavior.
- 1925: Samuel Abraham Goudsmit (1902-1978) and George Eugene Uhlenbeck (1900-1988) proposed the existence of spin. Spin is a property of particles, such as the electron, analogous to the angular momentum of a turning object.
- 1926: Werner Heisenberg (1901-1976) and Paul Dirac (1902-1984) proposed the exchange interaction to explain magnetic order. This interaction, of quantum mechanical origin, acts between electrons.
- 1927: Clinton Joseph Davisson (1881-1958) and Lester Halbert Germer (1896-1971) demonstrate that electrons have wave properties.
- 1931: First electron microscope made by E. Ruska (1906-1988).
- 1932: Louis Néel (1904-2000) suggests the existence of antiferromagnetism, a type of magnetism in which a sample has two lattices with antiparallel magnetizations of the same modulus, resulting in zero total magnetization.
- 1935: *Magnetophon*: the first magnetic tape recorder.
- 1951: First commercial computer: UNIVAC 1.
- 1953: Medical ultrasonography.
- 1956: The magnetic hard disk was introduced by IBM.
- 1956: First commercial video magnetic recorder.
- 1963: Magnetic compact cassette introduced.
- 1969: Creation of ARPANET, a network connecting four universities.
- 1971: First e-mail sent.
- 1972: IBM shipped the first commercial flexible disk drive (8 in).
- Late 1970s: The first personal computers appear.
- 1971: Magnetic resonance imaging.
- 1976: First commercial PET scanner.
- 1980: First commercial MRI scanner.
- 1981: The first precision map of Earth's magnetic field was made by the MAGSAT satellite.

- 1981: The Bitnet network, connecting scientific and educational institutions.
- 1982: The first optical disk (CD).
- 1984: The IBM 3480 tape drive was announced, with a storage capacity of 200 MB.
- 1991: The SSD (Solid-State Drive) was designed.
- 1991: WWW was created at CERN (European Organization for Nuclear Research, Geneva).

2000 AD

- 2001: Factoring of the number 15 using quantum computation.
- 2005: Perpendicular Magnetic Recording (PMR) of HDDs becomes dominant.
- 2013: Shingled Magnetic Recording (SMR) introduced.
- 2017: Two-Dimensional Magnetic Recording (TDMR) introduced.
- 2020: Energy-assisted (EAMR) magnetic HDs introduced.

Glossary

Antiferromagnet A type of substance that contains two lattices with *magnetizations* of the same modulus but opposite directions, leading to zero total *magnetization.*

Coercivity A measure of the sample's "magnetic hardness." Its value is given by the intensity of the magnetic field applied in a direction opposite to that of the magnetization of a sample, which is necessary to cancel this magnetization.

Compass A simple instrument formed of a magnetized needle that can turn, aligning itself with the direction of Earth's *magnetic field.*

Declination, magnetic The angle between the direction of the *magnetic field* on the surface of the Earth and the direction of the meridian, or the true North-South direction.

Exchange A phenomenon of quantum mechanical origin that induces an alignment of the electron *magnetic moments*, e.g., in a *ferromagnet.*

Hard magnetic material Material that has a large value of *coercivity* and may be used for the manufacture of permanent magnets.

Lodestone Earlier designation of a naturally occurring magnetic mineral; magnetized *magnetite*, or *magnetite* that presents a *magnetic moment.*

Magnetic field A property of space modified in the neighborhood of a magnet that leads, e.g., to the appearance of a torque acting on a *compass* needle or (when inhomogeneous), to a force on a magnetic object.

Magnetic moment Property of a particle or of an atom that measures its contribution to the magnetization.

The magnetic moment of the atoms The contribution of each atom to the *magnetization.*

Magnetite A naturally occurring magnetic mineral, an iron oxide of formula Fe_3O_4.

A. P. Guimarães, *A Longstanding Attraction*, https://doi.org/10.1007/978-3-032-02006-2

Magnetization A measure of the degree of magnetic order of a body, given by the total *magnetic moment* divided by the volume.

Order, magnetic State of magnetic materials that present *magnetic moments* arranged regularly, e.g., in parallel, as in a *ferromagnet*.

Paramagnet A type of substance that contains *magnetic moments* that are not ordered; a magnet attracts a paramagnet.

Index

A. P. Guimarães, *A Longstanding Attraction*, https://doi.org/10.1007/978-3-032-02006-2

MIX
Papier aus verantwortungsvollen Quellen
Paper from responsible sources
FSC® C105338

If you have any concerns about our products,
you can contact us on
ProductSafety@springernature.com

In case Publisher is established outside the EU,
the EU authorized representative is:
Springer Nature Customer Service Center GmbH
Europaplatz 3, 69115 Heidelberg, Germany

Printed by Libri Plureos GmbH
in Hamburg, Germany